Earth, Air and Water

RESOURCES AND ENVIRONMENT IN THE LATE 20th CENTURY

I.G. SIMMONS

Professor of Geography, University of Durham

Edward Arnold
A member of the Hodder Headline Group
LONDON   NEW YORK   MELBOURNE   AUCKLAND

© 1991 I. G. Simmons

First published in Great Britain 1991
Reprinted 1992, 1993

Distributed in the USA by Routledge, Chapman and Hall, Inc.
29 West 35th Street, New York, NY 10001

British Library Cataloguing-in-Publication Data
  Simmons, I. G.
  Earth, air and water: Resources and environment
  in the late twentieth century.
  I. Title
  333.7
  ISBN 0-340-52415-4

Typeset in 10/11 pt English Times
by Colset Private Limited, Singapore
Printed and bound in Great Britain for Edward Arnold,
a division of Hodder Headline PLC, Mill Road,
Dunton Green, Sevenoaks, Kent TN13 2YA
by Butler & Tanner Ltd, Frome and London

Contents

Preface

In the 1980s concern over 'environment' was again much in the public eye. A series of world conferences of both politicians and scientists endorsed its importance, especially where the likelihood of global climatic instabilities was concerned. All of these movements, however, treated the immediate factors likely to produce instability in a more or less isolated fashion: very often where 'pollutants' were concerned it was simply the production of wastes that were of interest.

When, therefore, Edward Arnold decided (and I agreed with them) that my book *The Ecology of Natural Resources* was too out of date factually and conceptually to merit a third edition, I welcomed this chance to replace it with an entirely new book which dealt with the same kinds of topics as the earlier volume but from an even wider perspective. So I have tried to deal with whole resource systems from their origins in raw materials right through to the disposal of wastes, which acknowledges that many of the 'environmental problems' which we identify have their origins in the search for, and use of, resources. Like so many Earth systems, the resource-environment interaction is often a whole and in trying to understand it we have to break it down into component parts and we lose some of the reality; but we have to start somewhere or else lapse into apathy.

The wider view has resulted in the word 'ecology' being dropped from the title but not from the conception, if we allow that rather over-worked word to apply to the study of the relation of organisms (including humanity) between themselves and their environments. Indeed the title is profoundly ecological. It is an adaptation of the last line of Thomas Hardy's poem 'Proud Songsters', in which he says that the year-old song-birds we hear were, only twelve months ago,

> . . . only particles of grain
> And earth, and air, and rain.

Since 'rain' was thought to be a little narrow in scope, we have substituted 'water'. But I have also tried to make more explicit, for example, the importance of resource and environmental economics, and also that of various behavioural traits of humans, including the widespread desire to tell others what they ought to be doing. Apart from a small flourish at the end, though, I have stayed within a materialist framework. I doubt this is the whole story but that needs yet another book.

So I hope this volume will be useful to a new generation of undergraduates in

Geography, Environmental Studies/Sciences and similar courses who found *ENR* congenial enough to keep it in print for 16 years and occasionally to send me on holiday on the proceeds. Mostly they were English-readers but translations also brought me comments (though not many royalty payments) from many parts of the world. I am grateful to all those who commented constructively on that book in reviews (Professor David Pearce of UCL comes especially to mind), in letters and at lectures.

This book has been a product of the last couple of years while I was in an intermission between two bouts of Chairmanship of my current department. Happily, my colleagues (and especially my Chairman, Professor John Dewdney) wanted very little of my administration 'talents' during that interval and so I had time to develop a new course on Environmental Management (which bears some relationship to this volume but is not as it were isomeric) and engage in a serious dialogue with a word-processor. This book is the first of mine to appear that went straight from brain to screen and I shall be interested to know if those commentators who in the past have been kind enough to praise my prose think that it has suffered in course of the use of the new technology.

In the production of this volume, a number of people have been especially helpful: Elizabeth Pearson edited onto disk many pages of additions and amendments to the text and Joan Andrews masterminded the final version of the print-out. The linework is a product of the over-stretched Drawing Office of this Department and I am especially grateful to Arthur Corner and his team (notably David Hume) who worked on them during 1990 and 1991. Mr P. Puvanarajan helped with other aspects of the linework. As always, the staff of the Inter-Library Loan section of the University Library were very helpful. In a slightly less direct way I owe thanks along with many of my colleagues to Dr Ian Shennan for encouraging us to use PCs in much of our work and thus release time for writing. Thank you all very much.

The Ecology of Natural Resources was dedicated to my parents. My father, alas, died in 1983 but my mother is still with us and still enjoying life: so this book is for her, with much love.

I. G. Simmons
Durham
March 1991

Units and terminology

Metric units are preferred throughout but occasionally imperial units are used in a historical or customary context. In the case of energy, the units of the original source are used provided that this was either calories or joules. In this case, equivalents can be calculated: to convert Joules to calories multiply by 0.239; to convert calories to Joules multiply by 4.184. A rough figure for human metabolism is an energy throughput of 3000 kcal/day. Multiples used quite often are k (10^3); M (10^6); G(10^9) and T(10^{12}).

In the case of characterising the income and development levels of nations and peoples, the coarse division of DC and LDC has been used informally where the context makes more accurate description unnecessary. Where a more refined classification is needed than the World Bank division of HIE . . . LIE is used: the categories are set out in Part 3, Chapter 1. Other acronyms are explained at point of first use unless they are so common that no explanation is needed.

Using the book

This type of book can have two main uses: to tell a story and to be a source of reference material. In order to facilitate the former, I have decided not to spatter the text with references and so students wanting suggestions for further work should look to the end of each part of the book where they will find a selection on Further Reading. I have tried to select material which is accessible to most students with access to a reasonable HE-level library but I am aware of the limitations which such collections face at present in many places. However, it is all material which I have personally seen recently.

The book contains quite a lot of numbers which will go out of date quickly. Unless there are traumatic dislocations such as continuing warfare in the regions of the major oil producers, we can assume that these numbers will be of the same order for the next few years. However, I recommend using the yearly production of the World Resources Institute, *World Resources 1990/91* is the latest, published by Basic Books, New York and distributed in Europe through OUP. Not only are the basic data tabulated there, but informative essays are presented as well.

Part I

RESOURCES AND ENVIRONMENT:
The Linkages

1

Introduction

Models of complexity

After outlining the way in which we can be linked to the material world around us, this section considers three case studies on the use of the resources of the planet, and the problems which are posed. We then draw some general lessons from these examples, (a) about the ways in which we can label our ways of studying them, and (b) about the general models of humanity and nature which are implicit in our studies. We end with asking how real are the models and the explanations which we give of them, a rather basic question.

The connectedness of humans

Each individual human requires certain inputs in order to survive, grow and reproduce. An adult has daily metabolic needs in the order of 3000 kcal of energy, 60g of protein, vitamins, and 2 litres of water. These, together with several other basic human requirements (such as shelter, clothing and access to services such as medicine) call for materials which are garnered from the lithosphere and biosphere. After these and other resources have been used, any wastes are led off back into water, land and the atmosphere. As a generalisation, therefore, we can say that humans extract material resources from their environments. Think of the kind of picture sometimes printed of a golf or tennis player whose club or racquet has a light on it which, when photographed, shows the trajectory of their swing; now imagine that every molecule of resource that we have used in the last month could be similarly tagged for its pathway from the environment to your body and out again: the foodstuffs and the fuels, for example (Plate 1.1). We can begin to see the complexity of the interaction of human societies and their global surroundings, especially in industrial countries. Even in developing nations the chances of the streaks of light ending at the boundary of the village lands are now very small.

Plate 1.1   Streaks of light emphasise the linkage within a city and in particular the way in which it uses energy. The main railway station at left, symbolises the external links of the city, its people, energy and matter. Cairo, Egypt, 1986.

Three case studies

What follows is a description of three examples of resource use, followed outwards from the human society into their environment in search of resources and in the creation of wastes. One is concerned mostly with a low income, another with the context of richer, industrial countries, and there is also a mid-way example, where characteristics of both can be seen. For all, we shall be concerned to trace the outward links of the resources uses described.

Woodfuel in the drier tropics

Today, only about 8% of the world's energy consumption by human societies is of wood. Yet in developing countries some 2/5 of the fuel supply is from biomass: about 1.5×10^{10} people cook or warm themselves with wood and in countries like Tanzania, 96% of the population are wood consumers. This means that in some nations a large part of the energy available to people can only come from wood, with supplementary biomass such as leaves, straw and dung where these are available. This proportion is as high as 90% in Burkina Faso, for example, and 80% in Nigeria, 93% in Malawi, 74% in the Sudan, 71% in Kenya and 50% in India. In Sri Lanka, woodfuel accounts for 45% of the fuel for industry and 84% of household fuel. Looked at on a world scale, it seems that the highest proportions of dependence are in the semi-arid developing countries and in mountainous regions like the Himalaya.

Demand for wood in developing countries averages about 1.5 t dry wood/cap/yr and, given rises in demand from growing populations and from urbanisation, shortages are becoming apparent. (The effect of growing populations is obvious; town dwellers exert an extra effect because they prefer charcoal and one unit of charcoal requires two units of wood for its production.) At the individual level, some women in rural Upper Volta now spend 4½ hr/day looking for wood; at the national level, in 1980 Sudan had a woodfuel 'gap' of 32 million m³; this is estimated to rise to 127 million m³ in the year 2000 (Plate 1.2). Shortages bring about a series of adaptations. In one rural African example these have been observed to be, in order of incidence, (1) that women have to spend much more time searching for fuel, (2) then that crop residues are added to the fuel repertoire, (3) then men start to join the hunt, (4) carts are brought in to help bring in the wood, (5) dung is sought out, (6) wood and other fuels are purchased rather than collected, and eventually (7) either fewer meals are cooked or a shift is made to different foods that need less cooking. Fuel efficiency is dependent on technology: a simple open fire is 6–8% efficient. Stoves can be 15–30% efficient but they are less portable and require maintenance skills that may not traditionally be available.

The transformations wrought by this search for wood or an easily available substitute are sometimes obvious, at other times less so. The spread of tree-cutting in the landscape may be a sign of the intensity of the search, although evidence from some places suggests that land clearance for agriculture is often more important, and that the trees are sold as a by-product of that process. Degenerate woodland, with most of the bigger trees taken out, is another result of the hunt for fuel; this land in turn may be then fully cleared for

Plate 1.2 In the Sahel zone of Africa, children carry wood fuel for cooking. They may have spent several hours in this task; note the generally bare nature of the settlement area but the preservation of a few trees.

agriculture. Rural people in general effect less ecological change in the search for wood: for them it is most likely to come as the by-product of some other land use. Towns, on the other hand, may bring about more thorough clearance because of the pulling power of their money and because urban dwellers prefer charcoal. In the semi-arid zones, both groups combine to produce a landscape with fewer trees in it and where those left have a lower chance of regeneration.

The scale of response to the supply shortages can be put in context by the figure that the Sahel zone of Africa needs a planting-up rate of 150,000 ha/yr to meet its fuelwood needs but is currently achieving 3000 ha/yr. A consensus among scholars seems to be that fuelwood needs to be produced in the setting of land devoted to other end uses as well as fuel, so that the growers are not monoculturalists. However, where a traditional forestry department exists, then its talents can often best be harnessed by the institution of peri-urban plantations, as has happened in units of 800–1400 ha around the cities of Malawi. Eucalyptus and pines are the most common trees in these circumstances. In more remote rural areas, new configurations for growing trees are being either introduced or revived. Intercropping is one of these: the presence of *Acacia albicans* among crops in the Sahel zone leads to higher crop yields. A systematic version of this using maize and *Leucaena* is called alley cropping. In tree plantations it may be possible to grow crops alongside the young trees, affording them protection and weed suppression. More wood can also be produced by persuading farmers to use trees for new purposes, from which fuelwood can be a by-product; soil improvement (using nitrogen-fixing species), fodder production, and pest management (in Africa, 1600 trees with pesticidal properties have been identified) are examples.

The outward linkages of the fuelwood process do not appear to be very strong. One connection is contained in the list of responses to shortages given above. If wood becomes scarce then crop wastes and dung may be used for fuel, thus diverting them from their use as fertilisers; in India, 10% of energy use is from burning cow dung. The crop either then goes without or has to have expensive chemical fertiliser. Another connection is simple: pay more for wood, have less money for food. More widely, the woodfuel supply and price networks interact with those for kerosene (paraffin), which has the potential to be a major substitute in urban areas. Given the price of oil and the shortage of foreign exchange in most of the developing world, (Sri Lanka for instance spends 40% of its export earnings on buying oil), it appears that only the already better-off are likely to make use of kerosene and thereby reduce a dependence on wood which is as strong as that of the developed nations on oil. Apart from a few gaseous contributions to the atmosphere, this interaction with the world of oil prices is the furthest-reaching of the light-links in our imaginary metaphor of the golfer and his swing.

Broad conclusions can be drawn. The first is that tackling the shortfall problem has to be done in the context of the indigenous rural economy and has therefore to be evolutionary rather than revolutionary; second, that if new technology can be made successful under local conditions it can raise the efficiency of the combustion process by a considerable factor; and third that the people whose lives are most restricted by low access to energy are generally the women.

Gravel working in Great Britain

In developed countries, each citizen uses or has used on their behalf about 3½ tonnes of stone, sand and gravel every year, (152×10^6 t/yr in Great Britain in the 1980s: the largest single extractive industry) which is largely used as aggregate. Since it is mostly locked up into structures such as concrete roads, foundations, and buildings, it can only be re-used as fill or as hardcore. Continued development therefore calls for new supplies of the original materials. Gravel and sand are however high bulk-low value resources and must be found as near as possible to their sites of use: a 48 km truck ride doubles the price at the pit.

Locational constraints are several: geology (especially Pleistocene history) determines in an absolute fashion where the deposits are to be found; costs and environmental amenity determine where they are to be worked. Thus the valley gravels of major rivers are often found to be major areas of gravel and sand working, (a) because the deposits are found there and (b) because these areas are the locations of most urban-industrial expansion. Such pits comprise about 66 percent of all the quarrying and open-cast mining in England and Wales at any one time: there are likely to be 1500 producing pits on a workday, extracting $28-37 \times 10^3$ m^3/ha from the low-level resources. To approach London Heathrow airport from the west down the Kennet and Thames valleys is to see a sequence of water bodies which result from sand and gravel working.

The transformations wrought by this type of mineral working are greatest in the river valleys since the materials are extracted from 4–6 metres below the surface, which means that the pit is effectively flooded from the start. The material is commonly extracted by a dredge and fed to a washing and grading plant, which is the only treatment it gets before use. Road transport means a heavy stream of trucks leaving the screening plan, with consequent noise, damage to the surface of rural roads, and the accumulation of fines by the roadside. The former land surface is reduced to a small fringe of solid land within the extraction parcel and a few transient water-birds are the only wildlife (Plate 1.3).

In most industrial nations, however, permission to extract gravel carries with it strict

Plate 1.3 Gravel extraction and sorting in southern England. Note the input of energy needed for the screening machinery. In the background a virtual desert of bare land awaits landscaping and restoration.

conditions about restoration. One course of action is to fill in the pits with e.g. domestic solid wastes. This is unpopular since the filling phase creates all kinds of nuisance and because any toxins in the wastes are likely to find their way quickly into groundwater. Inert rubble or fly-ash from power stations are the preferred materials for filling, which then creates land for housing, industry or recreation. Equally popular is the option to make the lake into a recreational resource, and given the demand near most western cities for water-sports this is often taken up; about 40% of the pits in south-east England have followed this course. The banks are usually landscaped and planted with grass and trees. Buildings are erected if required. Clubhouses, for example, are needed for sailing, water-skiing and windsurfing but not for coarse angling. If angling is to be the main use then a programme of stocking and possibly also of near-shore vegetation management is needed; some clubs also undertake (probably unwisely) control of predator species such as pike (*Esox esox*). A gravel contractor may also seek to enhance his public image by landscaping part of the pit so that its major use is as a nature reserve: to this end, shelving edges to the water body (rather than the steep drops favoured by the fishermen) may be created as well as some small islands. In the absence of influences other than perhaps some tree-planting, the bare areas soon get colonised with a variety of ruderal species, which then are progressively succeeded by shrub and woodland species on the dry land. The lake edge develops a succession of swamp and open water habitats and the whole ecosystem is usually highly attractive to both resident and migratory bird populations. These have the advantage of being close to population centres and thus provide a wildlife resource within easy reach of many inhabitants, who can indulge in 'twitching' within easy reach of their homes, thus lowering the demand for fossil fuels.

We often think of the resource linkages of the developed nations as being wide open and on a wide scale but here is one example of a relatively small compass. Only the energy needed to extract and transport the materials comes usually from a great distance: otherwise all the light-streaks are concentrated into the city and its environs.

The tropical forests

In say 1820, there were about 1600 million ha of tropical moist forests in the world, with distribution as in atlas maps. The demands created by the modern world upon these ecological systems are very varied. The population of the countries in which they occur have long used these forests as an environment for shifting cultivation, and for hunting (Plate 1.4). Selective logging has also been a long-term feature. Now, many areas are cleared with permanent cultivation or ranching in mind, though abandonment of the pasturelands is frequent after maybe six years. Some of this is government-sponsored settlement schemes, others the spontaneous movement of people to acquire some land. Clearance will usually mean that all the timber can be sold and there is a strong market: over half of Japan's demand for paper pulp is satisfied from tropical hardwoods. The demand for beef on world markets is also strong: a large hamburger chain may serve 9×10^9 hamburgers/yr, representing 300,000 head of cattle, many of which might come from pastures on converted tropical forest lands. The result has been the depletion of the forest area: in the years 1981–85, some $10–11 \times 10^6$ ha/yr were lost, about half each to shifting cultivation and to permanent loss. This included the conversion of *ca* 4×10^6 ha of virgin tropical forests to a secondary state. In these conditions perhaps only 10–20% of the timber is extracted but another 30–50% is damaged beyond usefulness. Yet these forests can offer a broad array of industrial materials from the 100,000 plant species which occur there. Essential and edible oils, genetic variety for domesticated crops and medicines are some examples. Perhaps only 1% of TMF species have been investigated for their usefulness.

Plate 1.4 In all the tropical forest zones, the trees are being removed at a rate which causes concern. Small-area attrition for shifting agriculture is one of the processes involved as here in northern Sumatra, Indonesia.

At present, to get to the trees and to get the cattle out, highways are built: 60% of the cleared land in Amazonia in the 1970s was from road-building, only 18% from peasant agriculture. The ecology of the forests is depleted by other forms of demand: in Germany an ocelot coat will fetch $40,000; and the skins of 2000 birds of paradise were exported from Indonesia in 1978. The same country between 1971–76 exported 91,000 macaques, mostly for medical research. In the case of Brazil, many native people have lost their livelihoods through development of the forest lands.

Here then we have a major transformation of parts of the earth's surface in the cause of seeking resources, just as the deciduous forests of Europe were changed to farmland in the medieval period, and the temperate grasslands ploughed up in the 19th century. In the case of the tropical forests the regression from their earlier distribution comes out at a world average (for the 1980s) of 42%, which comprises Africa at 52%, Latin America at 37% and Asia at 42%. The annual loss in recent years has been of the order of 15 million ha/yr (about the area of England and Wales), a rate of 30 ha/minute, and 6 million ha/yr of this total has been the lowland moist forests such as are found in the Amazon Basin. About 0.5–0.6% of the forests disappeared every year in the late 1970s. The ecosystems which take their place include tropical grasslands, secondary forest and scrub, savanna, short-cycle pine plantations, and inundated areas where dams have been built for power projects. (An area the size of Montana has been flooded in Amazonia.) Some land areas are not used once the forest is gone since there is no use for the consequent grasslands of species like *Imperata*, or stands of the herbaceous *Eupatorium* or the shrub *Lantana*. All this new vegetation is much less diverse than the legendary species richness of the tropical forests with possibly 400 tree species per hectare, compared with 15 in the temperate zones, and with 80% of the world's insect species being found in these forests. Their water relations are also altered once clearance takes place. Much less water is held by the forest ecosystem in its root zone: runoff is faster and contains more silt. There is more water at ground level, too, since evaporation and transpiration from the tree canopy is greatly diminished. Rainfall seems to be recycled from west to east across Amazonia but this may be less likely if it is diverted from evapotranspiration to the rivers. The solar energy taken up in evaporating the water is now released to heat the air, with the potential for climatic change.

The theme of climatic change links the changes in these forests to wider atmospheric concerns such as the 'greenhouse effect', at least 50% of which is due to enhanced CO_2 levels in the stratosphere. Part of this rise is due to the burning of fossil fuels (5×10^9 t/yr), and the burning of wood from the deforested areas ($3–6 \times 10^9$ t/yr). This latter comes from, for example, the 121,000 km^2 of forest and savanna burnt in Amazonia south of the Equator in 1989. Of this, 40% was rain forest. Equal in quantity and significance to this however is the diminished uptake of carbon from the atmosphere into agriculture and other non-forest land cover types. These areas simply do not sequester carbon at the same rate as fast-growing trees such as were found in the tropical forests. Some types of agriculture also contribute to the quantities of gases in the atmosphere which enhance solar warming: the 'greenhouse gases'. This is especially true of *padi* rice (not a factor in this example) and of domestic cattle, which produce methane, a greenhouse gas 21 times as effective as CO_2 in producing warming. Of the 485×10^{12} g of methane added to the atmosphere each year, 350×10^{12} come from human-directed activity, and 80×10^{12} g of that subtotal from domestic cattle. Thus the extension of cattle ranching in the tropics is part of the global 'greenhouse' problem.

If we imagine the connections of these processes to the rest of the world in terms of the golfer's lights, we will see that they stretch very far, and that they are not all very minor streaks: some indeed seem to be flashing red. One outcome of these processes has been to generate an international effort to persuade certain key nations to reverse their develop-ment policies and to protect their forests. In the pursuance of this, some international agencies have changed their loan and grant conditions, and agreed to forgo debts and NGOs have purchased areas of forest as reserves. Not least, in Brazil environmental activists have been killed by established landowners fearful of losing profits from ranching cattle. All of these impacts on the forests (commercial logging, ranching and small-scale cultivation) also amount to an unsustainable pattern of use.

Ways of studying complexity

Each of the accounts we have just read contains statements of particular types. These are grounded in the ways we have of tackling very complex matters such as society-resource-environment relationships. The basic way is one of reducing the complexity by adopting the already worked out methods of a particular intellectual discipline and also perhaps by restricting the boundaries of the explanation in time and space, to reduce its scope.

So, the first set of statements which we can discern is grounded in a scientific approach to the processes and problems that are described. We notice that the biology of the tropical forests, of the trees of the Sahel, and of the fish populations of abandoned gravel pits are all relevant to our discussions. In fact, it is usually the interaction between the populations of an organism, its physical environment, and the stresses produced by human activities that is the core of any study here. The branch of biology which deals with these connectivities is called *ecology*. Within it, the concept which draws together the living and non-living components of an area (at any scale from a drop of water to the entire planet) in a way which emphasises the dynamics of their interaction is called an *ecosystem*. Modern studies of ecosystems usually measure for example the changes in populations of the organisms which comprise it, the flows of energy from the sun through the living things and back out to space, and the flows of materials such as carbon or potassium between storage pools in e.g. the sea, the atmosphere or on land. So although the trees are the focus of any account of resource use in, for instance Amazonia, the whole ecosystem which binds together the trees, other plants, the soil, the runoff water, the rivers and their silt content, and even the heat and water balances of the regional atmosphere ought to be considered, for to affect one part (and especially the trees which function as a *key* element in the system) is very likely to affect them all. Our light-streak visualisation would work very well here: ecologists use radioactive particles in very much that way. Ecology is first and foremost a science and therefore aims in the end to provide a body of knowledge which is predictive. This would be very useful for resource studies since it might tell us the effects of changes which we bring about but because there are so many variables, ecology is not yet as successful in its predictive capabilities as for example physics.

Clearly, though, there are other forces at work in our examples. One major category of statements is about who can pay the price of the various items of resource that are available. The word 'demand' is used as shorthand for the powers of a would-be purchaser whose wants pull the material to him or her for usage. This idea applies alike to the customer in Megaburgers, the building contractor who is constructing a new branch of Megaburgers and needs concrete for the foundations, and the regional manager of Megaburgers who wants somewhere to sail the boat he can afford to buy because the company operates a profit-sharing scheme. What, then, we have here is the vocabulary of *economics*, which studies the way in which societies arrange the distribution of goods which are in limited supply, which is nearly everything except the air and even that in heavily polluted places. Economics is concerned with looking at the costs and benefits of a flow of a resource from the natural environment to its final disposal. In its classical form, it suggests that unless at every stage in the process the benefits exceed the costs and therefore somebody makes a profit, then the resource will not be developed. A band of light therefore comes to an abrupt end in our imaginary picture, since somebody cannot afford the price: a resource use is 'uneconomic'. One drawback with this view is that not everything can be easily priced: the development of a gravel pit may remove the last examples of water-meadows in a region, or diminish the number of pollarded willows that are the nesting-place of owls: how can their value (if we assume for the moment that both history and natural history have a value) be put in the balance sheet alongside the selling price of sand and gravel? Attempts have been made in capitalist economics but the outcomes are often disputed. Some economics is structured on different lines: to produce the greatest good for the greatest number of people, for example, or in order to combat the stratification of society into classes based on income. In all these, economists hope that their discipline is predictive, although as with ecology (with which it shares the same Greek root *oikos*, a household) certainties are hard to come by, apart from death and taxes. Ecology and economics do not easily interface with one another, though many attempts based for example on price, or energy flow, have been made to develop a common language.

There is a third way of studying all these human-environment interfaces. This is to regard them as examples of human behaviour, just as a zoologist would study animal behaviour, or *ethology*. So we observe the outcomes of many individual psychologies when we see poor people deciding to move to Amazonia and farming in a way which may give them a short-term income but poor prospects in the long-term because they may have been misled about the nature of environmental constraints. We note also the behaviour of the men in the Sahel who are reluctant to get involved in the search for woodfuel but will do so when their dinner is threatened, and the way in which the inhabitants living in a village near a proposed gravel pit will try very hard to prevent the development happening in their back gardens. Most of these behaviours exhibit in some fashion the exercise of power (like the African men or the middle-class house-owners) or the lack of it, like the poor farmers of Brazil. In terms of our streaks of light, some of them will terminate because an individual will choose not to get involved in a resource use, or will extend outwards since the reverse has taken place.

Human psychology is a wondrously variable thing, and its study reflects some of this plasticity. For our purposes, though, three types of approach will for the moment suffice. The first is to treat humans and the way they collectively assess and behave towards their resources and their surroundings in a scientific manner. This is called *behaviourism*. Counting can be done and classifications made and even predictions can be embarked upon. The second is the obverse. The world is seen as centred upon the individual and so studies focus on the way a person perceives the world adding immediate sensory responses to knowledge, history, beliefs, fears, income and other facets of his or her life. Regularities of such constructions can be counted and described, but the mainstay is the response of the individual in constructing a life-world which necessarily includes environment and resources. This way of studying is called *phenomenology*. Lastly, we notice that humans are often telling others how to behave: there is said to be a 'right' way and a 'wrong' way to use resources and to affect the environment, for example; in the developed world we think it is wrong for Brazil to clear large areas of rain forest and even worse for ranchers to have people killed who do not share their evaluation of the resource use of the forests. This outlook is called *normative* and, with the other two, contributes to a set of lenses through which we can view ourselves and the rest of the world.

Models of humanity, resources and environment

Behind these ways of investigation lie certain assumptions about the way the world is, and our place in it. These assumptions are often a part of our cultural inheritance which is taken for granted and not seriously questioned. We need to note their existence in case they provide us with alternative ways of proceeding when our evaluations of resources and environment suggest that we may be running into trouble.

The last sentence of the section above contains one such assumption when it states, 'can view ourselves and the rest of the world'. Here it is implied that *Homo sapiens* is different from the rest of our known environment and is qualitatively superior to it. This is an essential ingredient of what is called the Western *worldview*, which may be likened to a pair of contact lenses we wear and forget. Since the 19th century, the Western worldview has become dominant over the globe. It starts, as said above, with the centrality of humans: our view is quite literally *anthropocentric*. The environment (sometimes called *nature* in other texts and a synonym we can use here) can then be regarded as a set of resources for human use: it has no more important purpose than that. Our recent history also leads us all to expect that material progress will continue even if it is unevenly distributed, although we may disagree about the response we make to such inequalities. Where there are difficulties, then we look to science and technology to provide a way out, rather than necessarily changing our own behaviour. We reject therefore any suggestion that nature sets constraints to our activities: we want to have control over it. This attitude is usually summarised as being *instrumental*, since non-human nature is viewed as an instrument solely directed at fulfilling our own purposes.

Yet there are no reasons why this worldview should be the only one. In other cultural traditions, or in the 'alternative' thinking of the West, there are other choices. For example, there is no reason why humanity should not co-exist on equal terms with other animals and with plants: they could have just as high a value as that which we accord to ourselves. Nature *is* in its own right therefore and not merely as a set of materials for human use and enjoyment. This attitude accords an *intrinsic* value to the non-human world. Equally, there are reasons why there might be limits to growth of material possessions and comfort, or of cities, or of human population itself. And it is open to us to consider that science and technology might sometimes be a source of problems rather that solutions. Finally, we ought to consider the possibility that nature does exert finite limits upon all our activities and that if we are not satisfied with that, then we can try and colonise other planets.

Since the western worldview is strongly influenced by the philosophy and methods of science, it also tends to model the world in a *reductionistic* way. That is, to break down the system into component sub-systems. If the answer is not apparent at the ecosystem level, study the component organisms; if they do not yield it, then look at their organs, or tissues or cells, or molecules . . . as far as is needed or is feasible. This view, too, has its other side, which holds that a whole, such as the ecological functioning of the planet, is more than the sum of its parts. Breakdown into components, from this *holistic* viewpoint, inevitably causes a loss of reality. An analogy is sometimes drawn with human personality: a holist argues that personality emerges from all the biochemical processes of the brain but cannot be explained by them. In terms of the light patterns of our tennis player, the whole pattern is the story; looking at only a few of the streaks will never tell us how to improve the player's service.

These models all have a long history. A view of humanity and nature as essentially one entity has been a characteristic of some modes of Asian thought for millenia. The notion that human societies are determined by the nature of their physical environments dates back at least to Classical Greece. Here were formulated the kind of ideas that lie at the heart of contrasts between north and south in many European countries: that southerners are passionate but lazy and northerners controlled and hard-working. Basically, it is held to be the climate that is at the root of this. Unpopular since the early 20th century, this *determinism* has emerged since the 1960s on a global scale with the idea of limits to growth: that somewhere there must be physical limits to the quantity of resources that can be produced, the amount of wastes that can be absorbed by environmental systems, and hence there are limits to the number of humans that the planet can support. This worldview is usually labelled *environmentalism*.

The creation by man of his own world, in his own image, within the world of nature is also found in Antiquity. From very early recorded times, humans have found that the planet did not provide an ideal habitat for them. Some parts were too wet, others too dry, too mountainous or too heavily forested for easy access to a livelihood. Some of these features could not be changed, but others were amenable to alteration. Swamps could be drained, forests replaced by agricultural lands, and predators upon cattle killed. In western Christendom, for instance, this could be presented as holy work: either as polishing up a Creation which was incomplete or as trying to return it to an Edenic state, since wastes and fearsome forests were tangible evidence of the Fall. Generally, the outlook of the Benedictines who were willing to drain, reclaim and deforest prevailed over the Franciscans whose founder preached to birds and told wolves not to be predatory. Nevertheless, every time we change a part of our environments with good intent (what we will probably call 'wise use'), we are acting in the tradition of creating a second world within the world of nature, and in space vehicles we take this to one extreme.

Are we realists?

At the very beginning we have to be aware of the findings of philosophy that warn us of the risky enterprise on which we embark when we say anything is 'real' or 'true'. Here is not the place to explore this problem, for we shall adopt a scientific standpoint which acts as if the experience of our senses, as confirmed by data-gathering and by the same results from independent observers, actually produced truth. At the back of our minds, we might just remember (and occasionally recall) that everything including science is a construction of our minds and there might be something about even the minds of professors which is by definition incapable of recognising the whole truth. For the time being, however, the viewpoint of philosophical *idealism* which hold that there is no world outside our heads, will be set aside in favour of a *realism* which accepts that those things out there continue to have an existence whether we perceive them or not.

2

History of Resource Use

Population, resources and environment in the past

No apology is needed for a consideration of the past; we need to know where we have been in order to have datum lines against which to assess the present and the future, while always acknowledging that discontinuities make simple forecasts impossible. Thus in this chapter we shall chronicle the growth of the human population in the last few thousand years, and match this to an account of the accompanying growth in resource use. The combined effect of these upon the natural environment will also be studied in a historical perspective.

The growth of the human population

If we believe one very early written source of quantitative information about the human population, then there were a few years of zero growth when the absolute number was two. However, Adam and Eve were nothing if not innovative and so began the steady climb of the demographic curve to its present (1990) level of *ca* 5255 million. But looking beyond the myth to the more scientific evidence, it seems as if the genus *Homo* evolved in southern Africa about 4 million years ago, and that our present species *Homo sapiens* is a product of the last 40,000 years. It too began in Africa, but quickly fanned out from that continent, especially in the wake of the retreating ice sheets at the end of the Pleistocene, about 10,000 years before the present. It is human history in this latest (Holocene) period which concerns us here.

Two aspects of this growth may be mentioned at this point. The first is that every human can be considered as requiring two sorts of resources. The first we can call *metabolic* and are those which are simply needed for survival. Each additional person added one unit of these of his demands upon the planet as a support system. The second we can call *cultural* and derive from the culture: our metabolic need for water may be 2 litres/day but if we live in a modern western city then we are more likely to use 600 litres/day. So every extra western urbanite places a very much larger demand upon the earth than a rural inhabitant of a simple culture. Even when the population of the earth was much smaller than now, and living entirely in a pre-industrial economy, the latter demands caused some concern among writers: as early as classical Greece, some thought that the earth was full of people and could support no more. We particularly associate this view with T. R. Malthus

(1766–1834), an English clergyman who has given us the term *Malthusianism* for a concern about the relations of population and resources which thinks of the future with misgivings. It is an attitude which still waxes and wanes today.

The course of human population growth can be divided into four major phases, the first of which comprises the Late Pleistocene and early Holocene, when hunting and gathering was the sole form of human economy. During these periods (culturally named the Upper Palaeolithic and Mesolithic), the supportable densities of people averaged at about 2–3 people per 100 km², i.e. 2–3 people in an area nowadays bounded in London at its four corners by Charing Cross, West Ham, Walthamstow and Wood Green. One estimate for global population once the major ice-covering of the northern hemisphere had melted, is of 4 million in 10,000 BC.

Soon thereafter we can distinguish the beginning of the second major phase, which was made possible firstly by the movement and expansion of people into areas hitherto unoccupied in Oceania and the Americas, and secondly by the widespread adoption of farming. The use of domesticated plants and animals produced much more surplus energy per unit area per worker than hunting and gathering and so higher densities of humans could be supported on suitable lands. This Neolithic revolution allowed growth rates of 100% per century in the last three millenia BC, so a population of 100 million is postulated for 500 BC, with AD1 seeing a total of 170 million. Thereafter, growth rates seem to have slowed up to give an absolute level of 265 million by 1000 AD (Fig 1.1).

The next phase was led by Europe and China and although one of considerable growth overall, suffered some checks: the Mongol invasion of China in the early 13th century resulted in 35 million deaths and destroyed much of the agricultural structure, and in

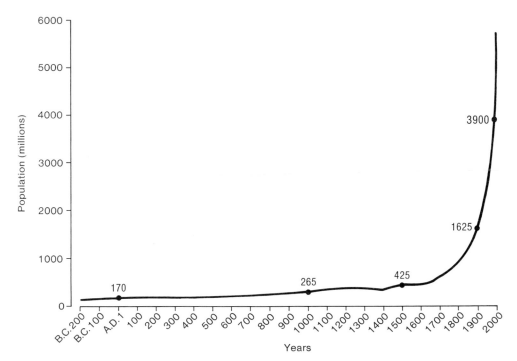

Fig. 1.1   Growth of the world population from 200 BC to the present. Source: C. McEvedy and R. Jones, *Atlas of world population history*. Harmondsworth: Penguin Books, 1978.

Europe the plague brought down the population from 80 million to 60 million. It seems as if in Europe at least the population rose to the carrying capacity of the agricultural systems of the time in about 1300 AD. On a world scale, the total in 1400 AD was about 350 million, having been 10 million higher a century earlier, reinforcing the idea that some form of Malthusian check was imposing an upper limit.

Any such restraint was lifted in the modern period, when technology, colonisation, industrial development, new food-producing techniques, scientific knowledge, and modern medicine have all combined to lift the limits pertaining in the medieval era. Growth rates seemed to take off in the 16th–17th centuries but especially after 1700 AD, again fuelled especially by China and Europe. In the 19th century, for example, Europe's population gained by 135%, of whom 20% emigrated to other lands. This phase is different from the others in the sense that it is largely attributable to a decline in mortality and especially to a reduction of deaths from infectious diseases. In this trend it is likely that better nutrition and higher standards of hygiene played a part, though the exact mix of causes is still argued about by historians. Whatever the explanations, a world population of 350 million in 1400 AD became 700 million in the last quarter of the 18th century, 900 million in 1800, 1625 million in 1900 and 5000 million in 1989. Of these people, less than 1% lived in Oceania, 20% in the Americas, 12% in Africa, 16% in Europe and the rest in Asia.

The current position is one of two major world types of population: those which are

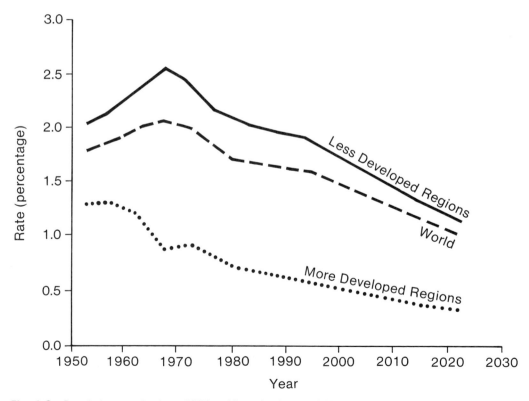

Fig. 1.2 Population trends since 1950, with projections to 2020 AD. The lines for various types of regions are for annual rate of growth in percentages. Source: *World resources 1988–89*. NY: Basic Books, 1988.

Table 1.1   Historical growth rates of world population, per cent per year

	World	All industrial	All developing
1650–1750	0.34	0.33	0.34
1750–1800	0.50	0.62	0.47
1800–1850	0.43	0.83	0.31
1850–1900	0.68	1.05	0.53
1900–1950	0.98	0.75	1.09
1950–1960	1.83	1.26	2.07
1970	1.90		
1975	1.64		
1986	1.70	0.6	2.0
1989	1.70	0.6	2.0

Sources: J.D. Durand, 'A brief history', in Q. Standford, (ed) *The world's population. problems of growth.* Toronto: OUP, 1972, 14–21; WRI/IIED, *World Resources 1987.* New York: Basic Books 1987, and subsequent annual editions.

growing slowly or are nearly stationary, and those which are growing quickly (Fig 1.2). In the first category, the declines in mortality which fed the rapid rates of growth (up to *ca* 1% per year) in the 19th century have been followed by reductions in the birth rate; in the second group there are still high birth rates even though mortality is down and so annual growth rates average at 2% (Table 1.1). The slow growth areas comprise Europe, USSR, Japan and Oceania (19% of world population) together with North America (5.5%). Faster growing regions comprise Africa (11.5%), Latin America (8.4 per cent) and Asia except Japan (55.8%). In total, the industrialised nations have about 24.4% of world population and the developing countries 75.6%. Some projections for the year 2000 suggest that the proportions then will be of the order of 21% and 79% respectively (Fig 1.3).

The low-growth regions share certain demographic characteristics. The major lands

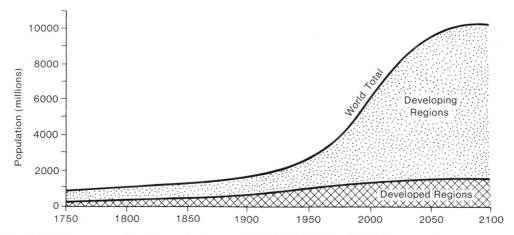

Fig. 1.3   The course of world population growth 1750–2100 AD divided between the two major types of economies. (1 billion = 1000 million) Source: *World resources 1988–89.* NY: Basic Books, 1988.

involved are Eastern and Western Europe, the USSR, Australia and New Zealand, North America and east Asia, including China as a developing country with a growth rate of only 1%. This group had a population of 2,322 million in 1986 and its average rate of increase was 0.8% per annum: a few nations like West Germany were actually losing population but all were at or below the China level of 1% per year. All show a combination of rising levels of material benefits and falling levels of fertility, though the social milieu of the latter is different in e.g. Western Europe from that in China. The reasons for having fewer children involve such factors as the costs in time and money of raising them, the buffering provided against old age by the state or by private pensions, thus reducing reliance upon children, and the revaluation of the role of women in society. The results include aging populations, with the 65 + cohort growing the fastest. In developed nations, some 13% of the population is 64 + years old and 22% is <15 yr; in the less developed, the figures are 4% and 39 per cent respectively. In fact, in some developed nations, much of the population growth results from immigration and the group with the fastest natural increase is that of recent immigrants, though they later adapt to the national custom in this respect.

In terms of demand for resources and impact upon environment, these slow-growing populations still exert the heaviest pressures because of their economy and affluence: their effects are heavily weighted by cultural demand as well as the high absolute numbers. However, they are also the places with the highest awareness of the long-term consequences of what they are doing and the sources of new thinking. Whether their demands will lessen in the face of ageing populations (fewer demands for ski-slopes for example) ought to be the subject of studies of life-cycle changes in demands for resources such as energy, space and materials.

The faster-growing populations of the world are mostly in the less-developed and poorer regions (Plate 1.5). These are found in S. E. Asia, Latin America, the Indian sub-continent, the Middle East and Africa, and within these broad categories there are countries with very large populations and very high rates of growth, such as India (785 million, 2.3% per annum), Indonesia (168 million, 2.1% per annum) and Brazil (143 million, 2.3% per annum). Africa has the highest set of population growth rates in the world and within that continent, the sub-Saharan regions are the highest, such as Kenya at 4.2% per cent per annum (δ = 17 yr). These fast-growing areas have a world average increase rate of 2.5% per cent per annum (δ = 28 yr) of their current 2700 million population, to which they add about 66 million people per year.

These then are lands of high birth rates and declining incomes. Per capita incomes fell between 1980–86 for example by 28% in Nigeria, by 16% in the Phillipines, by 21% in Argentina and 11% in Peru. Further, most of them have very heavy burdens of external debt so that much land has to be devoted to export crops to service those loans. Thus grain production per capita 1970–85 fell by 17–25% in several sub-Saharan countries, by 24% in Peru and by *ca* 50% in Mozambique and Haiti, emphasising the role of political troubles. These populations are still subject to high levels of infectious and parasitic diseases, and cancer is increasing where smoking is becoming more prevalent. These patterns produce heavy impacts on local resources such as fuelwood, soils and water: some of this pressure is simply metabolic, deriving from the sheer numbers of people and the rates of increase. Many of them, therefore, are dependent upon outside aid for part of their nutrition: thus they make resource demands in other parts of the world as do their richer cousins.

In a world which adds to its humans numbers by about 192,000 per day (about half the population of Iceland), which comes to 70 million per annum (i.e. the population of Austria or Zambia), there is much interest in projections of future numbers (Table 1.2). Most of these are educated extrapolations of present trends and hence subject to

Plate 1.5   This crowded street stands symbol for the high rates of population growth in developing nations (note the age distribution) and of the attractiveness of the cities. Note also the gender distribution. Calcutta, 1988.

inaccuracy: most past predictions have turned out to be wrong. Nevertheless, increasing sophistication of modelling is providing better guides to the range of future numbers, and in this field the work of the United Nations is outstanding. The factors which will affect future levels of population the most are mainly those affecting people in the Third World since that is probably where 95% of the population growth in the next 100 years will take place. Thus the critical variables will be the cultural changes in those countries which affect the age of marriage, the importance of children in the labour force, the costs of raising children, the prevalence of breast-feeding, the social, educational and occupational status of women, and the degree of urbanisation. The flow of knowledge and materials to limit family size is also important, but secondary since the decision to be informed and then to use appropriate ways of spacing children is the most important. Modernisation of LDCs may initially raise fertility, for better nutrition and hygiene by themselves may well lead to the survival of more infants. The success of family planning in bringing down birth rates when conditions for acceptance are favourable is beyond doubt: Costa Rica for example reduced its BR from 47/1000 in 1961 to 28.5/1000 in 1985, and the world rate fell from 32.2 in 1970 to 26.0 in 1985.

The UN Medium Variant Projection is one of the most commonly used guides to future numbers. It suggests a doubling of the world's population by 2100, with Africa having the highest growth rates, which will increase the number of people on that continent from 555 million to 2.6×10^9 by 2100. The biggest absolute increase will come in Asia, from 2.7 to 4.9×10^9 by 2100. This Variant sees replacement fertility level [RFL] (i.e. zero growth in

Table 1.2 Summary of world population data

	Estimated population size ($\times 10^3$)		Annual growth rate (%)	
	1985	2000	1980–85	2000–05
World	4 842 048	6 127 117	1.67	1.39
N. America	400 802	488 073	1.38	1.33
S. America	268 825	359 581	2.24	1.63
Asia	2 824 008	3 543 693	1.73	1.22
Europe	492 009	513 110	0.33	0.18
USSR	278 373	314 818	0.95	0.69
Oceania	24 820	30 403	1.5	1.19

Source: United Nations

terms of natural increase) being achieved by 2035, which would bring about a stabilisation of world population by 2100 at a level of 10.2×10^9. (The World Bank goes for the same year but at a level 1×10^9 higher.) If the replacement fertility level of 2.1 children per family is delayed then the absolute number of population is increased and the year of achieving stability is put back, and vice-versa. Thus if RFL came 20 years earlier then there would be 2.2×10^9 fewer people; if it were 20 years later, then 2.8×10^9 more. The difference is today's world total.

Projections should be looked at with a sceptical eye. Not only must the best science go into them but they ought also to be the product of accurate and frequent censuses, which is rarely the case. Projections may be as much politics as demography and we need to be wary of optimistic prognoses based on land area per person ('everybody in the world could get onto the Isle of Wight') as if land were homogenous in its productive capacity; and of pessimistic ones which simply extrapolate from some convenient figure like 2% per annum and end up with shipping the surplus off to space. Equally misleading are the type of value judgements inherent in calling people an 'ultimate resource' as if they existed in a vacuum and as if they might not be equally happy at lower densities of population. Nonetheless we need projections since they enable us to measure the likely demand for certain basic resources (given various assumptions) and to see the range of possible environmental outcomes of meeting those demands.

A history of resource use

The world has changed a great deal since 10 000 BC and much of that change has been at human hands. The complexity of those alterations is immense: look out of a window on a 30-minute journey in most parts of the world and try to identify all the ways in which what you see is created by humans in search of resources rather than provided by a pristine nature. For the present purpose, we are going to assume that the power to get a living from our natural surroundings and to change them is conferred by the success of human cultures in getting access to energy and that the levels of energy use are a surrogate for levels of resource use.

For humans, energy has two functions: the first is the metabolic use which requires an intake of energy in order to fuel life itself and an average intake of perhaps 2500 kilo-calories per day is needed to keep a healthy adult alive and moving. The second function is unique to our species and is the use of energy outside our bodies ('extra-somatic energy') in the form of technology of many kinds, many of which gain us access to resources and enable us to change our surroundings, sometimes deliberately and sometimes by accident. This gives us the cultural resources of which we have spoken earlier.

Hunter-gatherers

The energy sources of this group of people are those of recent solar origin. The most important is that from food which represents recently fixed solar energy in chemical form. To get this energy, humans expend their own in the form of for example running and stalking, digging pits, setting nets, using spears, blowpipes and the bow, digging roots, and collecting fruits and seeds. What is critical is that every collector brings home a surplus of energy to feed those who cannot do it for themselves like the very young and the elderly (Plate 1.6). To this should be added if possible a reserve against a season of shortage. A key aid in this equation was the discovery of the control of fire, since this helped hunters to direct game towards waiting men; but beyond that it soon must have been apparent that vegetation could be manipulated by repeated firing to the point where the new plant community was more productive of human food than the old, either because the plants were different or because they harboured more game.

Agriculturalists

The ability to control fire was not lost after the coming of agriculture but it ceased to be so central. Instead, the garnering of an energy surplus was enhanced by the use of domesticated animals and plants. This meant that the sites of growth, the control of times of cropping and the essential characteristics of the plants and animals used were under human direction. Instead of being gathered from hundreds of square kilometres, plants were

Plate 1.6   Hunter-gatherers existed in more or less traditional economies well into the 20th century. The Canadian North was one such place: this group of Inuit are shown near their houses and very little impact of the industrialised world is visible. *The picture dates from c. 1920*

Plate 1.7 The immense environmental impact of pre-industrial agriculture is demonstrated in this picture of rice terraces in Bali in 1978. Powered machinery may now be used but construction and maintenance for hundreds of years has been solar-powered as has the sophisticated water-control mechanism.

grown in fields next to the settlements; animals were coralled nearby instead of chased all over the landscape. So adequate nutrition and some surplus could be gained from hectares not square kilometres, although the input energy levels from the cultivators were often very high, in the form of ploughing, sowing, weeding, fencing, harvesting, threshing and processing, for instance (Plate 1.7). The harnessing of ancillary energy sources such as those of children and domestic animals to do labour, and wind or water to process food in mills became an important part of the agricultural era.

Industrialism

Before the 19th century, the great storehouses of concentrated potential energy in the form of coal, oil and natural gas were not totally unknown. Coal was used to warm the soldiers on Hadrian's Wall and natural gas piped through bamboo stems lit streets in a few medieval Chinese cities. But the realisation of the potential of the steam engine, coupled perhaps with the rising price of both wood and labour, made the new process of smelting iron with coke the key to a complete revolution in human-environment relationships. Here then is a shift to much older energy (millions of years in fact) and furthermore to energy in a form that is not being laid down today. Results were first the metals revolution which gave rise to a technology based on iron and steel and then the chemicals revolution (Plate 1.8), to which we may possibly add the plastics revolution. Think also of all the other new materials as well, like carbon fibre and superconductors. These have given humans the power to

Plate 1.8   Early industrial plant: part of the Floridsdorf refinery in Austria in the 1880s. A throughput of 2900 t/yr of crude oil was converted into paraffin wax (for candles), lamp oil and lubricating oil.

make over the natural surroundings on a scale never before contemplated. The consequences have been enormous and so complex as to defy easy categorisation. One change has been in the speed and ubiquity of communications: starting with the steamship and ending for the moment at least with satellite television and the fax machine (Plate 1.9); another has been the setting up of the factory system for the production of manufactured goods and the consequent attraction of a labour force, hence the explosive growth of cities in nations undergoing industrialisation. Industrialisation is not confined to industry, but is now interlinked with most forms of economic activity: agriculture is often heavily subsidised by commercial fuels, for example, and the mass tourism industry relies totally on them for cheap transport. Even climbing Everest needs an industrial base to produce the oxygen cylinders and the hi-tech clothing.

The nuclear age

Alone of all these stages, we can attach definite dates to the evolution of the processes. For this, let us choose 1942 when Enrico Fermi engineered the first controlled chain fission reaction, thus opening the way for both civilian and military nuclear power. Beyond the immediate horizon lies the possibility of generating energy from atomic fusion, in which an easily available isotope of hydrogen would be the chief fuel. Just appearing over it are the so-called 'alternative' sources of energy which tap the renewable processes of nature such as wind, waves, tides and sunlight. Their environmental impact in anything other than the visual sense has yet to be evaluated.

A history of environmental change at human hands

This has been a continuous process since the Pleistocene but we shall break it up for convenience into the same phases as used above since access to energy confers the ability to change the environment (whether deliberately or accidentally) just as it confers the ability to create and to use resources.

Plate 1.9 Global telecommunications are made possible by a certain amount of fossil-fuel energy and a lot of scientific and technological knowledge. These telecom dishes in London can relay speech and images almost anywhere in the world via satellite links. The background building style rather sums up the 1980s, too.

Hunter-gatherers and environment

The acts of hunting and gathering by their very nature alter plant and animal populations. But the pressure exerted upon the populations is usually such that if the humans leave then the biota return to their previous levels. It is in the humans' interests, too, that their exploitation should never extirpate their resource supply, i.e. that they practise a form of management for sustained yield. Through history some groups seem to have had ways of making sure that they did not over-use plant and animal sources, whereas others did not, relying perhaps on their own very low population densities to bring about only minimal effects. An animal population under natural environmental stress may not be able to cope with human hunting pressure and thus become extinct: this seems to have happened in several places throughout the world just at the end of the Pleistocene. Fire, too, can produce permanent transformations in marginal environments, where e.g. the removal of trees by burning alters the water balance and peat bog results.

Agriculture and environment

With the coming of the domestication of species, we can begin to pick out zones of greater intensification of environmental alteration. There is for instance the difference between shifting and permanent systems. Where an agricultural plot is abandoned when its yields decline, or herds are moved on quickly by nomads, then something like the original vegetation may become re-established. By contrast, some permanent agro-ecosystems may represent so much inherited effort that abandonment would be cultural as well as economic barbarity. Rice grown on flooded terraces (*padi*) is one example of this: some tropical hillslopes have been totally transformed by these constructions. Some systems may of course over-use the land and change it to a less productive state: land badly eroded by wind or gulleyed by water is one example. Heavy densities of grazing animals, too, may bring a pastureland to a near-desert condition. The energy surpluses of agriculture mean that not everybody has to engage in subsistence activity: thus other occupations are possible which in turn may produce different manipulations of ecological systems. An aristocracy, for instance, may hunt for pleasure and so may erect parks with walls or fences to keep in beasts for the chase and so protect that species against the extinction which herdsmen outside the fence greatly desire. The great may also require the construction of elaborate gardens or landscape parks, transforming either wild land or the productive land of lesser folk. Industry, too (though not on the scales of later times) was present in city and countryside: iron smelting and tanning in Europe for example both pushed the management of forests in the direction of coppicing to produce a steady flow of small poles to feed the furnaces with charcoal in the one case and to yield bark (that of oak was a good source of tannic acid) in the other. Industry and the towns produced wastes, themselves in turn agents of environmental changes. Inland fish populations were altered by the intensity of harvesting but as yet those of the oceans were still economically in the hunting phase and not permanently affected so far as we know.

Industrialism and the environment

The advent of the use of fossil fuels gave societies many tools with which to affect their surroundings. The immense energy surpluses made available from coal, oil and natural gas could in part be channelled into getting at more and more resources in progressively less accessible places and so the imprint of industrial society spread. But it also became more intensive in its heartlands (such as Northern Europe, North America and Japan) and so at two scales, we can envisage the environmental consequences of industrialisation as a core and periphery model. Starting at the local level, we can imagine a turn-of-the century industrial plant such as a large coal mine or perhaps a coke works or steel mill. The local environmental impacts are dominated firstly by the change in land use, and the local air and water will be contaminated in plumes stretching downwind and downstream from the plant. This reaching-out of the effects of the plant is paralleled on the input side by the tentacular stretch of the growing conurbations for water: local upland valleys are likely to be submerged by reservoir construction.

If we extend this model to the whole planet, the developed areas of the world can be regarded as the plant or the conurbation, and the rest as the zones of (a) outreach for materials, and (b) sinks for wastes. One of the chief demands of the industrial zones was for food and so for example, many temperate grasslands were either ploughed up to provide cereals such as wheat or converted to ranching to provide meat. As motor vehicles took hold, the demand for rubber increased greatly and so tropical forests in for instance Malaya were converted to rubber plantations, with sales of the original timber an additional source of profit. Many other agricultural systems now found that growing for export was possible and so some changed their ecologies drastically by importing irrigation techniques, for example. Another great impact of industrialisation upon the planet's resources was in the seas where steam trawlers facilitated the catching of fish as never before, and a map of the dates when particular fish populations became uneconomic to utilise is a map of (a) proximity to early-industrialised areas and (b) dates starting in the 1880s. Even before 1918 the hunting of whales was getting much more effective, with the aid of steam-powered mother ships and the use of explosives in heading harpoons. The greater access to energy resources made available in the 19th century had its impact upon pleasure as well. The railways and steamships began the era of mass travel for pleasure of which today's inheritor is the package holiday to Spain or the Seychelles, with all the environmental change that comes in its wake.

The nuclear age

The scenarios for 'nuclear winter' pointed out the devastation to our environment that would be caused by even a modest 5000MT exchange of thermonuclear weapons. The incident at Chernobyl and other accidents have also made us aware of the potential for contamination in the civil sphere; we need also to remember that planned releases of radioactivity also take place during the nuclear fuel cycle: the sediments of the Irish Sea for example have the highest concentrations in the world of such materials outside direct human control. The apparent alternatives to nuclear and fossil fuel power, such as wind, wave, solar and geothermal energy also have their environmental impacts: some are visual, others involve chemicals such as sulphur in high concentrations. We have yet to assess the environmental implications of the latest developments in biotechnology and telecommunications but there seems no doubt that their potential is very great, in terms both of actual impacts, of e.g. engineered plants and other organisms, and the potential manifestations of the rapid diffusion of large quantities of information.

Nobody need doubt, therefore, that the world we inherit is largely man-made at the land surface, and to some extent below and above that surface as well, and at the fringes of the oceans. Yet the basic biogeochemical flows and cycles, of energy, carbon, and nitrogen for instance are still the basis of the human life-support system; though we may have greatly modified nature, we have by no means been able to replace it.

3

Resource systems

The search for a framework

We need a way of describing all those flows of energy and matter: one that encompasses the number of humans and their various cultures, the flow and use of materials and the biophysical context ('the environment') in which it all takes place. Preferably, it will be a way which can cope with changes through time as well and although 'realist' in the sense that we shall assume that it is there even we are not observing it, a good description will not be blind to the information we have about the ways in which individual humans achieve the perception and cognition of their surroundings.

The notion of systems

The classic way in which such a task is nowadays undertaken is through the idea of the system. At its simplest, the system can be thought of as a set of components together with the relationships between them. It thus contains from the start the notion of a dynamic interaction between parts but is not confined to any one type of component: in ecology the notion of an *ecosystem* has been very fruitful in the study of the interactions between populations of plants and animals and their non-living environment, for example; in anthropology the same framework has pointed to critical points in the way particular groups of people have wrested a living from their surroundings.

A simple model of a system is often presented in the form of a flow diagram, and in Fig 1.4 an attempt is made to construct a generalised resource system which takes in the criteria outlined just above, notably the fact that humans live in two worlds: the ecological/physical and the psychological. Central to the model is the human population, which straddles the two spheres of (a) the physical and ecological, and (b) the psychological. Humans sense the environment via a series of neurophysiological processes (perception) which are informed by all the various strands of our culture. The two together can be called cognition. Some of the elements of culture are learned in a formal way, such as science or economics, others are acquired as part of an unquestioned worldview. Our history, religion, and national prejudices are shared parts of this skein; our own hopes and fears might be more personal parts of it. Thus through the compound lenses of this cognition and perception we evaluate the biophysical environment in a mental way: what

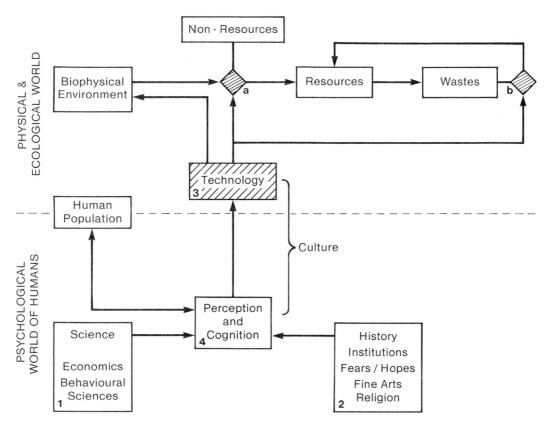

Fig. 1.4 Humans in two worlds: that of nature and that of the mind. The diamonds represent the choice-points of human societies, and the numbers refer to: (1) Formal 'constructions' of the environment by appropriate branches of formal disciplines; (2) Some other factors which are constituents of people's perception and cognition; (3) Technology as a mediating force: in effect, applied energy; (4) Perception is usually a neurophysiological process, cognition the cultural whole or *gestalt*.

can we do? what must we do? what ought we not to do? are being asked subliminally. What in fact we actually do is governed to a great extent by the tools that we append to our bodies in order to move the components of the environment (e.g. by digging) or to move ourselves (by a pair of shoes or a vehicle). Summed up, we gain access to environmental resources by means of technology. At the same time, technology informs us about what is in the environment: satellite remote sensing is be a good example. Culture (of which technology is a part) can thus divide the world into resources and non-resources. Resources can in this instance mean both material objects like iron ore or cotton or total environments like a National Park; non-resources are usually materials outside the range of a culture: like iron ore in Minnesota was said to be of no use to the native Amerind population since they knew nothing about smelting. The list of non-resources gets smaller and smaller as the range (in all senses of the word) of western culture increases.

Resources themselves are then brought into the human fold and used. The use may be consumptive in the sense that the resource is transformed into the course of its use into a form which no longer has utility for that culture in that form. A worn-out motor car, for example, may possibly be of value in bits (as spares and as scrap metal) but not *qua* car. Food becomes heat, water and solid excreta, of which the latter in some cultures is cycled back to the land but which in others is eventually burned to get shot of it. So here we have the notion of wastes: the by- and end-products of resource use. Once again, culture comes into play in evaluating these wastes. In some culture very little is classified thus: we have all seen the pictures of poor people picking over the trash heaps in cities in search of anything that can be sold (Plate 2.33). By contrast, rich societies have a great problem knowing what to do with all kinds of disjecta, from household garbage to the most toxic products of the chemical industry.

Conventionally, wastes are either reclaimed for further use (recycling) or led off into the environment as a form of (we hope) ultimate disposal: land, sea and air are all used, in the hope that the concentrations will be so low as not to matter to anything living (sea and air) or that the material will effectively be sealed off from all other biological and physical processes, as in landfills and the use of old mines and quarries.

This all forms an interactive system in which most of the components can change their relationships. Cultures can alter and make a material that was formerly a non-resource into a valued item or, *vice-versa*, they may decide that not to use a place like Antarctica is better than to exploit any minerals and oil that may be discovered there. The whole is driven by the dynamic of the numbers of humans (metabolic demand for resources) and their material requirements which constitute the type of cultural demand in which one European consumes in a year enough commercial energy to last a Bangladeshi for 65 years. The trajectory of the system is irreversible: oil and coal once burned cannot be un-burned; people once having lived and consumed cannot be un-lived. Even renewable resources like water and organic materials from plants and animals have cycling times (like the residence periods of water in the various phases of the hydrological cycle or the reproductive cycle of living things) so that the time dimension of both the whole and the parts is critical to its functions at many scales.

The whole system or parts of it can also be subjected to the sorts of questions raised by our earlier discussion of viewpoints. We can ask about the ecology of the system or its sub-systems: is it stable? That is, does it fluctuate within predictable limits which do not threaten human societies e.g. by sudden floods or droughts, over-productions or famines? For the economics viewpoint we ask, is the system profitable in some way to those who carry out the operations to make the resource flows possible. Even though the notion of profit may be different in a CPE, somebody has to gain from such processes. Are we all gainers or does somebody or something also have to lose? Then we must ask, is the flow culturally acceptable: does it destroy something which is held in such high value that there is a strong movement for keeping it, or does it produce for example blue food? A stable and sustainable resource process will give positive answers to all these questions and they will be interrogations which we shall want to make of several systems as we come across them.

Characterising systems

The two main halves to Fig 1.4 also give us another way of looking at resource systems. This is to classify their study into two modes: the objective and the subjective. In this book, we shall mostly use a disinterested and 'scientific' mode of analysis which purports to look at each system from the outside and to evaluate it objectively and dispassionately in the light of all the data at our disposal. This is not actually possible, of course, but we can have this aim. The type of approach is perhaps best called *functional* since its major task is to see how the system appears to work, to find a way of describing that which is intelligible to as wide an audience as possible, and then to evaluate it against cultural criteria. It follows that this best suits the ecological and economic questions asked above, but is less easily applied to the cultural, though it may be possible to incorporate them as choice-points in the flows.

The other approach is the inverse: i.e. it is *reflexive*, starting with the life-world of every individual person and using that as the starting-point for studying resource flows. We look not at the global patterns of streaks of light which have coalesced into a general pattern, but at what emanates from one man or woman. The difficulties with this way of proceeding are that the experience of the one person may not be transferable to a wider audience (though some basic experiences such as hunger are presumably much the same) and that an individual's time horizons are likely to be rather limited compared with that which can be discussed by observers paid to take a long view. Nevertheless it does us all good to be reminded that the whole point of having access to resources is to allow people not merely to stay alive but to have access to education, good health, recreation, and affection and esteem, and we have no meters to measure the sum of these: success is seen in peoples' faces and heard in their spontaneous words. However, this book will largely adopt the functional approach.

Open and shut cases

The globe forms a closed system except for energy, plus a few bits of space-programme debris. Hence, all materials are confined to the planet and their flows are in theory knowable: though not of course in actuality. Energy is different since the planet receives radiation from the sun and re-radiates heat back to space: for these flows the planet is an open system. In general, we know the magnitudes of these flows and the types of wavelengths of energy involved: if we did not then we would have no way of understanding the processes involved in phenomena like the 'greenhouse effect' and the 'ozone holes'. This approach reminds us that there are overall physical limits to human activity. There is, if nothing else, the number of people who could be supported on the incident solar energy all transformed into plant material since we cannot eat sunlight. This number has no significance since much of the energy is incident on the sea, which is not a human habitat. Further, living at those densities and at the limits might be something to be avoided. But it is a powerful antidote to excessively Promethean thinking to have to remember that the planet is to all intents and purposes the only habitat we have or are likely to be able to afford.

At sub-global scales, though, it is equally true to say that all systems are open systems and capable of receiving and transmitting energy and materials. This was so before humanity began to exert much effect upon the world's biophysical systems and now is at an even greater level since with technology materials and energy can be transported about the globe in large quantities. Drawing boundaries of resource systems in spatial terms becomes ever more difficult when regions, nations and blocs begin to trade, when gaseous and aerosol wastes are in the upper atmosphere and in the Greenland ice, and when every small town has a Chinese restaurant. Just as important in breaking down autonomous spatial systems is the transfer of ideas, especially via telecommunications, and we have yet to see what satellite TV will produce in terms of demands for material resources fuelled by images from other cultures.

Further reading

General works

Dasmann, R. F. 1984: *Environmental conservation*. Chichester: Wiley, 5th edn.
Ehrlich, P. Ehrlich, A. and Holdren, J. P. 1977: *Ecoscience: population, resources, environment*. San Francisco: Freeman, 3rd edn.
Rees, J. 1990: *Natural resources. Allocation, economics and policy*. London: Routledge, 2nd edn.

Introduction

Tropical forests
Bunyard, P. 1985: 'World climate and tropical forest destruction', *The Ecologist* 15, 125–36.
The Ecologist 17 (4/5) 1987: 'Save the forests, save the planet', 129–204.
Jordan, C. F. (ed) 1987: *Amazonian rain forests. Ecosystem disturbance and recovery*. NY: Springer-Verlag Ecological Studies 60.
Kartawinata, K. and Vayda, A. P. 1984: 'Forest conversion in East Kalimantan: the activities and impacts of timber companies, shifting cultivators, migrant pepper farmers, and others', *in* F. Di Castri et al (eds) *Ecology in practice. Part I: Ecosystem management*. Paris & Dublin: UNESCO & Tycooly Press, 98–126.
Longman, K. A. and Jenik, J. 1987: *Tropical forest and its environment*. London: Longmans, 2nd edn.
Myers, N. 1983: 'Tropical forests: over-exploited and under-used?' *Forest Ecology and Management* 6, 59–79.
Repetto, R. Magrath, W. *et al*, 1989: *Wasting assets: natural resources in the national income accounts*. Washington DC: World Resources Institute.
WRI/IIED 1988: *World resources 1988–89*. New York: Basic Books.

Woodfuel in semi-arid lands
Eckholm, E. P. 1975: *The other energy crisis: firewood*. Washington DC: Worldwatch Paper no 1.
Foley, G. Moss, P. and Timberlake, L. 1984: *Stoves and trees*. London: IIED Earthscan Paperback.
Leach, G. and Mearns, R. 1988: *Beyond the woodfuel crisis. People, land and trees in Africa*. London: Earthscan Publications Ltd.
Munslow B. *et al* 1988: *The Fuelwood trap. A study of the SADCC region*. London: Earthscan Publications Ltd.
Smil, V. 1987: *Energy, food, environment. Realities, myths, options*. Oxford: Clarendon Press.
Smith. N. 1981: *Wood: an ancient fuel with a new future*. Washington DC: Worldwatch paper no 42.

Frameworks

Barbier, E. G. 1989: *Economics, natural resource scarcity and development. Conventional and alternative views*. London: Earthscan Publications.
Jansson, A-M. (ed) 1984: *Integration of economy and ecology – an outlook for the eighties*. Stockholm: Ask Laboratory.
Johnston, R. J. 1989: *Environmental problems: nature, economy and state*. London and New York: Belhaven Press.

Mooney, H.A. Vitousek, P.M. and Matson, P.A. 1987: 'Exchange of materials between terrestrial ecosystems and the atmosphere', *Science* 238, 926–31.

Pearce, D. Markandya, A. and Barbier, E.B. 1989: *Blueprint for a Green economy.* London: Earthscan Publications.

Pearce, D. and Turner, R.K. 1990: *Economics of natural resources and the environment.* Hemel Hempstead: Harvester Wheatsheaf.

Proops, J.L.R. 1989: 'Ecological economics: rationale and problem areas', *Ecological Economics* 1, 59–76.

History

Boyden, S. 1987: *Western civilization in biological perspective: patterns in biohistory.* Oxford: Clarendon Press.

Jones, E.L. 1981: *The European miracle: environments, economies and geopolitics in the history of Europe and Asia.* CUP.

Simmons, I.G. 1989: *Changing the face of the earth. Culture, environment, history.* Oxford: Basil Blackwell.

Part II

Resource systems

4

Energy: the binding resource

Energy's special role

Our discussions in Part I have shown how important energy can be conceptually in, for example, showing a structure in the whole of human history, and in formulating the idea of a resource system, where it can provide one of the links between the interacting components. In this chapter we turn to the more practical tasks of seeing how it works as the common element in many resource systems.

Energy in natural and human-based systems

If we imagine our streaks of light picking out only energy flows over the face of the Earth, then it will be immediately apparent that some derive solely from the natural fluxes, whereas others come in whole or in part from systems in which humans are involved. Energy flow can be measured therefore in both types of system and the magnitudes compared. This is one way of seeing the extent to which humans have succeeded in their search for a resource use pattern which frees them from some of the limits apparently imposed by nature. Table 2.1 shows measurable flows for a variety of systems, natural and human, averaged as energy per unit area (1 square metre) per day. What is striking is (a) that natural processes (Plate 2.1) dominate the upper levels completely, and (b) the highest human-directed level is the city, which is of course a concentration of energy and materials, with high densities of energy use and material confluence. Cities, though, are only made possible now by the availability of fossil fuels and nuclear electricity. It is worth noting that environmental hazards such as volcanoes and earthquakes are overall very small releases of energy but their impact is often great since the actual occurrences are confined to a small area. What we can infer from this Table (2.1) is a picture of a planet in which the biggest energy flows are still those provided by natural processes such as climate and photosynthesis but one in which human-directed energy consumption (of the non-renewable fossil fuels) is now *ca* 8 per cent of the renewable energy fixed as plant tissue (Net Primary Productivity). So while humanity seems overall quite a small feature, nevertheless by tapping concentrated forms of energy rather than sunlight and by concentrating

Table 2.1   Global mean energy flows for various natural and human-induced processes

Process or event	Energy flow (cal/m^{-3}/day)
Solar energy to earth	7000
Solar energy absorbed by earth	4900
Weather	100
Primary production by plants	7.8
Hurricanes	4
Tides	1.54
Animal respiration	0.65
Cities	0.45
Forest fire	0.3
Fossil fuel	0.11
Urban fire	0.065
War (non-nuclear)	0.05
Floods	0.04
Earthquakes	0.001
Volcanoes	0.0005

Source: J. F. Alexander, 'A global systems economy model' in R. A. Fazzolare and C. B. Smith (eds.) *Changing energy use futures*. NY., and Oxford: Pergamon Press, 1979, vol. 1, 443–56

Table 2.2   Growth in Population and Energy Use 1870–1986

Year	(1) World Population (millions)	(2) World use rate of industrial energy (terawatts)	(3) Per capita use (watts)	(4) Cumulative energy use since 1850 (terawatt–years)
1870	1300	0.2	153	3
1890	1500	0.5	333	10
1910	1700	1.1	647	25
1930	2000	2.0	1000	55
1950	2500	2.9	1160	100
1970	3600	7.1	1972	200
1986	5000	8.6	1720	328

Source: J. P. Holdren, 'Energy and the global predicament: some elements of a sensible strategy', in M. Levy and J. L. Robinson (eds) *Energy and agriculture: their interacting futures. Policy implications of global models*. Chur, Switzerland: Harwood Academic Press for U.N.U., 1984, 91–93. Additions from WRI 1988/89. Note: 1 terawatt (TW) = 10^{12} watts = 31.5×10^{18} J/yr = 14×10^6 bbls/day of oil.

activities into urban and industrial agglomerations, mankind can begin to approach the orders of magnitude of the biophysical processes of this planet.

The critical nature of access to energy resources in the various phases of human history can be summarised by Table 2.2 which shows the cumulative amounts of energy which have been consumed by societies in the course of their activities since 1870. We do not have data for earlier times but only the advent of fossil fuel use in the 18th and 19th centuries merits the term 'industrial energy' in Column 2. Column 3 sums up all the energy use until 1870 as 3 TW-years, whereas the rate per year was 8.6 in 1986 and the accumulated sum to date is 328 TW-yr.[1] This represents a formidable change in the quantity of resources flowing

1. A Terawatt is 1×10^{12} watts.

Plate 2.1 The natural energy flows of the planet include vulcanism. An aerial photo of an eruption in Hawaii in 1984 shows the local concentration of heat and molten rock. Yet on a global scale this is not a significant energy flow.

to human societies, for most of the goods and services which we use are provided with the aid of certain quantities of fossil fuel energy. This is called *embedded* or *embodied* energy and the idea is elaborated below. These global data are brought to a more individual level in Column 3, where the per person uses are set out: note (a) the steady rise throughout the 19th and 20th centuries to a peak of nearly 2 kw per head in 1970 (enough to run a 2-bar electric fire all day) but (b) after 1970, the rate of use has fallen back somewhat because of continuing population growth even though production has continued to rise, and remember (c) that even the 1.7 watts/cap is by no means evenly distributed, as we shall see later. Even at the 1986 figure, however, the energy available to each individual to enable her or him to get access to resources and to change their environments is 11 times that for 1870 (i.e the power levels have risen by over 10 in a century) and given that some individuals and societies have very low energy use levels, there must be some who consume a great deal.

The simple tabulation of levels of energy use is not the whole key to the resource question, needless to say. For each of us has a basic metabolic need for energy to stay alive and to reproduce: so much is biological. But because we formulate a world in our heads in which there exists motor cars, the city of Sydney and wines like Montrachet, there has to be energy available to devote to workers for Nissan, commuters to King's Cross and to labourers in the vineyards of Burgundy. Whether this comes from an agriculture so productive as to feed many beyond the farmer and his family or from oil that can be turned eventually into food and drink is subsidiary to the over-riding question: is there a surplus after the immediate producer has regained his or her input energy? Is there a surplus of energy on a remote farm in the tropics after the farmer and his family have eaten enough to be able to plough, hedge, ditch, sow and weed? Is there a surplus of energy after the oil company has explored, drilled, piped, refined the crude oil, and attempted to clear up after the last major spill in Arctic waters?

We may surmise that the production of energy surpluses has driven much of human material culture and perhaps has not been that remote from more metaphysical concerns as well. For instance, it seems unlikely that many hunter-gatherers had large surpluses and if they did they were not turned into material possessions, to judge by the archaeological record. Ethnological evidence, indeed, has suggested that a potential surplus might have been deliberately avoided by some hunter-gatherers simply by not 'working' (i.e. neither hunting nor gathering) all that much: in some places 2 days/week kept a whole group well fed. By contrast, the cities of Mesopotamia, Egypt and the Indus valley during the millenia before Christ could only thrive either if their inhabitants fed themselves (not impossible if there were intensively used plots nearby) or if the regional agriculture provided storable surpluses. (Here is the metaphysical interest, for any priestly élite would have wanted to ensure its own food supply without resorting to actual labour and so might well have invented a whole cosmos in which their presence was essential to the growth of the wheat or other staple food.) A reconstruction of the agriculture of the Nile Valley in the 18th century showed that an energy input of 9.61 GJ/ha/yr of animal and human labour returned 19.25 GJ/ha/yr of edible human food energy along with 62 animal feed days of clover and 122 feed days of straw.[2] Such an output would have provided, if the same relationships held good in much earlier times, for surplus wheat to be exported to the Roman Empire or for the diversion of labour to the building of pyramids. Other agricultural surpluses are chronicled by T. Bayliss-Smith: his data collected in Table 2.3 show the energy surplus from a variety of farming systems of relatively recent times. They tell us, for example, that the surplus from an English farm of the 1820s was similar to that from a simple farming system in Papua New Guinea in the 1970s; that collective farming in the USSR did not produce a great surplus (which connects with the shortages in that nation) and that the fossil-fuel backed agriculture of today does indeed tend towards Euro-mountains.

Once we have access to the rather concentrated energy available in fossil fuels, then the surplus is potentially much greater. Table 2.4 shows that hard coal (but not brown coal or lignite), oil and natural gas have an energy content well in excess of the wood which is the main fuel of a pre-industrial group. In turn, nuclear power is very much higher than the fossil fuels. Thus, any energy invested in developing fossil and nuclear fuels seems likely to yield a manyfold return. If for the moment we ignore the nuclear case (dealt with later), then Table 2.5 shows us the ratio of investment to return in energy technologies in the USA in the 1970s. The input side does not include refining, transport and labour, so is essentially the ratio at the mine-head, hydropower dam or other *in situ* piece of plant. The data are

2. A Gigawatt = 1×10^9 watts.

Table 2.3 Energy relations of some agricultural systems

System	Surplus energy (MJ/cap/day)
New Guinea, 1970s	2.3
England 1826	2.4
S. India	7.5
Moscow collective 1970's	4.1
England 1971	18.8

Source: T Bayliss-Smith, *The ecology of agricultural systems*.
Cambridge: CUP, 1982
Note: A surplus of 2.0 MJ cap/day is equivalent to
8474 Kcal/day, i.e. food for nearly 3 people.

Table 2.4 Energy content of fuels

Fuel	Value in J/kg
Coal (hard)	29×10^6
Coal (lignite)	12×10^6
Peat	4×10^6
Crude oil	43×10^6
Petrol (gasoline)	44×10^6
Natural gas	55×10^6
Charcoal	28×10^6
Wood	14×10^6
Dung	17×10^6
Bread	10×10^6
Fission (pure ^{235}U)	5.8×10^{11}
Fusion (pure ^{3}H)	3.3×10^{14}

Collected from various sources. The values for organic
materials are approximate since the composition is to some
extent variable. Note that most of the organic materials are in
the same order of magnitude but that these are then great
leaps into the yields of nuclear fuels, through the technology
of civilian fusion (as distinct from the hydrogen bomb) has not
yet been developed.

indicative, however, for oil and coal especially: prime the pump with one bucket of water
and at least 20 buckets come pouring out for further use (Plate 2.2). To anticipate slightly,
the data for fission reactors show a rather lower surplus (of the order of $\times$ 4) than fossil
fuels but their protagonists claim many other kinds of advantages for them. (The measure
of input to output is often designated as E_r where r = ratio).

If we are right in supposing that energy and society are so closely interwoven, then we
might expect that quite sweeping claims for its role in bringing about a particular type of
society are made. In this book we shall not investigate them closely but we might bear
certain of them in mind as we read of the more everyday concerns and processes. It seems,
for example, as if a resource surplus (and hence presumably an energy surplus) is necessary
for a society to have a relaxed form of government rather than an authoritarian one where
harsh decisions have to be made about allocation of scarce resources. One historian indeed
postulated that the shift from biological and climatic sources of energy to geological

Table 2.5　Energy return and investment* for some fuel supply technologies in the USA

Process		Energy return
Oil and gas 1970s		23.0
Coal (at mine head)	1950s	80.0
	1970s	30.0
Oil shale		0.7 – 13.3
Ethanol (ex plant residues)		2.6
Methanol (ex wood)		1.9
Solar flat plate collector		11.2
Liquid geothermal		4.0

* 'The ratio of gross fuel extracted to economic energy required to deliver the fuel to society in a useful form'. The measure presented here excludes however the energy costs of refining, transport, labour, government and environmental services.

Source: C. J. Cleveland, R. Costanza, C. A. S. Hall & R. Kaufman, 'Energy and the U.S. economy: a biophysical perspective', *Science* 225, 1984, 890–97.

sources was the dynamic behind the breakdown of feudalism. Taking this further, the control of energy flows in an industrial society might be seen as the key to the political power as well. When in 1990, Moscow wished to bring Lithuania to heel, its first reaction was to cut off oil and gas supplies to the rebellious republic. Likewise, there were strong assertions that the Gulf War which started in January 1991 was motivated as much by the desire of Western nations to maintain their oil supplies as it was by moral considerations. On the other hand, a less deterministic attitude can be taken by asserting that energy merely makes certain things possible and that it is the psychology of the society which determines whether or not a particular piece of technology is deployed. In this view, the possession of the motor car made sprawling suburbs possible but not inevitable. Here we enter upon a complex and long-running philosophical debate about whether technology (which we can see as largely applied energy) is controlled by society or has some autonomous quality once it is up and running: more than one invention has been likened to the story of the genie and the bottle, or to that of the Sorcerer's Apprentice. Some commentators liken energy companies, for example, to the fur-traders in North America who supplied the Indians with whisky and then when the hunters were firmly hooked on the hooch, started to screw up the number of pelts needed to exchange for a bottle of the water of life.

There seems little doubt that some forms of technology impose a template over human activity: nuclear power, for example, requires the acceptance of a high degree of security measures anywhere near the plant and materials in transit since the theft or escape of uranium or plutonium is fraught with disastrous consequences. Thus the power plants have armed guards even in nations where the police do not carry firearms. The computer will often determine the way in which information is gathered, irrespective of the relation of the format to reality: only the things that can be counted, count, it is said. It is worth noting that this analysis goes back as far as the 17th century: Francis Bacon's (1561-1626) *New Atlantis*, a kind of Utopian vision, had a board of guardians, one of whose jobs was to determine whether any new invention might be allowed into society or kept secret. So any radical change in the technology of resource use might sensibly be subject to some form of audit to try to determine its future consequences but in western societies today, the market is generally held to be competent to winnow out the harmful from the beneficial.

Plate 2.2   Drax power station in Yorkshire, England, 1988. Coal is the fuel. As well as heat and steam, carbon, sulphur and nitrogen compounds are emitted into the atmosphere. Some attempt to lessen the landscape impact has been made by using banks of soil.

Whatever the case, it can scarcely be denied that the whole western lifestyle (to which those who do not have it mostly aspire) is underlain by plentiful and relatively cheap supplies of energy. Making a few assumptions here and there, it took me about 5 minutes (before tax) in December 1990 to earn the price of 1 imperial gallon (4.5 litres) of petrol at UK prices.

The nature of energy

Energy is defined by physicists as the capacity to do work, which is done when an object is moved against a force. Energy is found on this planet in a variety of forms, some of which are immediately usable by human societies to do work, whereas others require transformations. The main basic forms present on Earth are:

- solar energy: the radiant energy from the sun, which reaches the surface of the Earth as heat and as light as at various wavelengths. It is the basic material for the most important form of

- chemical energy, which is solar energy fixed by plants during the process of photo-synthesis. This is present in recent form as wood, for example, and in fossil form as oil and coal. The net amount of chemical energy in plants per unit area per unit time is called Net Primary Productivity and requires water as well, which is the basis of

- hydropower, which is the energy contained in falling water. This is normally generated by the energy of the sun in evaporating water from whence it comes under the forces of gravitation and runs downhill to the seas. Gravity also causes the tides, in the course of which sea-water also runs downhill for part of the day. The land surface which channel all these flows also contain rocks which are the current source of

- nuclear power, which is released when the nucleus of a heavy atom such as uranium is fissioned. The raw material is uranium ore which is mined like many other rocks. Nuclear energy is also set free when the nuclei of very light atoms such as hydrogen are fused under conditions of extremely high temperature and pressure.

Human societies, however, cannot make machines or in other ways tap all these forms of energy in all the ways in which they are presented by the biophysics of the universe. Ingenuity, especially since the 17th century, has given us a number of ways of transforming these energy sources into usable forms (Fig 2.1). Thus solar energy may evaporate water, which rains out and is caught in an artificial lake, then falls and is used to generate electricity which then is transmitted by power lines to a machine which turns an axle and helps the manufacture of, say, paper. Oil is refined into petrol (gasoline) which is then oxidised to release gases whose pressure drives a piston that is linked to an axle upon which is mounted the chassis of a vehicle. All these processes have in common two facts:

- the law of the conservation of energy, which states that energy is never in fact lost, but merely transformed in form; and

- the second law of thermodynamics which states that energy is most likely to change from a concentrated to a dispersed form, in which condition it is unable to perform work. Chemical energy for example in our food, becomes heat which radiates into the atmosphere and can do no work, though it can warm a room for a while. This dispersed form is usually that of heat: it is a product of all the energy transformations that occur during this process of movement from concentrated to dispersed form.

It is therefore essential to think in terms of the whole energy flow of a resource process. All such operations eventually take concentrated energy and produce a dispersed energy which can do no work; this is called the production of entropy and is inextricable from life on this planet. We can however slow down the rate of its production and make sure that we do not spend energy on destructive things, like the 7% of the world's economy that is devoted to the military. Any proper audit of energy use, moreover, must calculate not only the daily energy throughput of a system, but the energy taken to construct the system and to dispose of it after its useful life. The energetics of automobiles thus should encompass not merely the fuel economy of the vehicle but the energy costs of building it, providing roads and parking, and disposing of its worn-out body. Every joule of concentrated energy spent on one process is, after all, not spent on another: every artillery weapon precludes the manufacture of another plough or the training of another agricultural extension worker; every horse to be fed from grassland prevents the growing of maize for human food.

Energy and resource systems

The examination of energy in the preceding sections makes it possible to think of resource systems in terms of their energy-flow characteristics, from those which are solar-powered on the one hand to those which are dependent upon complex technology powered by fossil and nuclear fuels (Table 2.6).

Solar-powered resource systems are present in those parts of the world which are subject

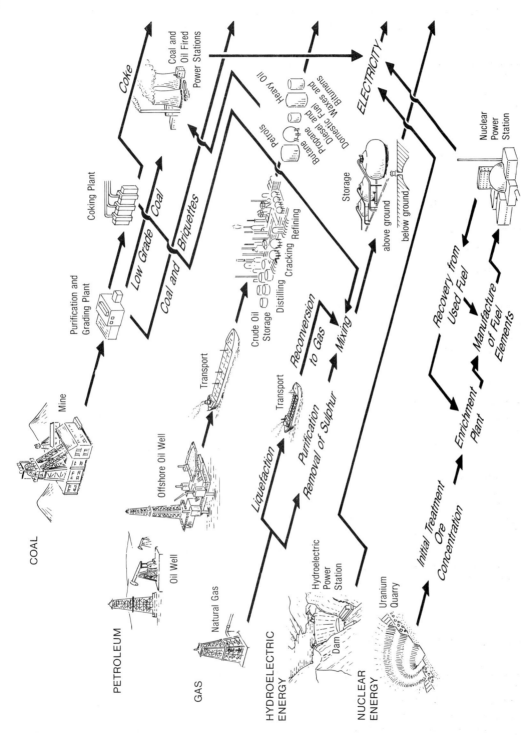

Fig. 2.1 Commercial energy production and conversions: the major sources and flows. Redrawn from J. T. Mullan *et al Energy resources*. London: Edward Arnold: 1983, 2nd edn.

Table 2.6   Energy Sources and Resource Systems

Energy Source	Type of economy		Notes
	Industrial economies	Low-industry economies	
Mainly solar energy, organic sources	Added energy = 100,000 kcal/cap/day	Added energy = 80,000 kcal/cap/day	Even in LIE's often some added FF energy e.g. as imported metal tools
	DHP with wood Woodstoves for domestic heating, luxury use Nature conservation Wilderness recreation	Hunting and gathering Fishing Pastoralism Water supply Wood & biomass fuels for space-heating and cooking DHP with wood Agriculture, esp. shifting agriculture Wildlife reserves Traditional medicine	Agriculture in LIE's continuously variable with next category
Mixed solar and fossil fuel/ nuclear sources	Added energy = 200,000 kcal/cap/day	Added energy = 100,000 kcal/cap/day	IE industry may have past-solar raw materials (drugs, tobacco, some textiles, paper) but use FF economy at most stages up- and downstream
	Agriculture & ranching Forestry and fisheries Outdoor recreation including hunting Land reclamation, waste disposal Natural fibres Water supply	Agriculture with chemicals & machines where used Pastoralism with vehicles Tourism Forestry Land 'development'	Even in LIE's tourism requires FF's for transportation, construction
Mainly fossil fuel/nuclear sources	Added energy = 400,000 kcal/cap/day	Added energy = 150,000 kcal/cap/day	Communications depend upon FF's to tie all the information flows together and to provide services in e.g. education, medicine finance
	Energy extraction and processing (mines, wells, refineries) Manufacturing based on metals, plastics Urban recreation & tourism Urban/industrial growth Land dereliction Education Medicine	Ore extraction and processing, refineries Toxic waste reception and storage Urban tourism Education, 'modern' medicine	
Systems with hi-tech renewables	Added energy = ?	Added energy = ?	Limited penetration since FF still cheap in IE's. Delivery of electricity to rural LIE's an important component of development
	Domestic lighting (windmills, small turbines) Farms using biogas Hitech solar for hot water HEP on industrial scale	Domestic heating & cooking Biogas HEP on industrial scale	

Note: This Table is not intended to contain exhaustive lists, nor that every cell is self-contained: intermediate states exist. The existence of electricity-based telecommunications ties almost everything together (except perhaps LIE solar-based systems) in terms of information flow.

to intermittent human presence almost all insufficient to cause anything but the most transient of changes to the environment. Thus remote mountain areas, most of Antarctica, the most arid desert regions, and the deep oceans may have on occasion some technology-based energy delivered to them but the energy flow at a given place is from the sun, either directly or via weather systems. But if a region is used for wilderness recreation, for example, people will be brought to its margins by engined vehicles. If another region is devoted solely to wildlife, then any management function will probably be technology-based: poachers hunted from the air, or helicopters used to count animal numbers, for instance.

Any added energy brought about by permanent human presence may still be at a low absolute level if it is solar-based. It may, though, be concentrated through channels such as domesticated plants and animals, water and wind mills, and so be capable of bringing about considerable environmental change especially when pursued for millenia. Pre-industrial agriculture, i.e cultivation and pastoralism as practised before the 19th century (and still to some extent found in LDCs today) is of this type, and today's organic farmers would like their system to be so, though this is difficult to achieve when it is embedded in economic systems with different energy relations. This category does not exclude industry powered by falling water or by wind or by wood as fuel, often in the form of charcoal. As with the solar-only category, human pleasure is a motivation in some of these systems: consider the intensity of energy applied to gardening during the centuries, for instance.

With the full-scale addition of fossil and, later, nuclear fuels the picture changes in most ways. The range of technological devices now made possible is very great and no attempt will be made to list them all. Note, however, that many are mobile: their energy source is so concentrated that they carry it with them, a revolutionary trend started with the steam locomotive and carried on into the farm tractor, the bulldozer and the dragline excavator. This access to highly concentrated energy sources has immensely increased the access of human societies to resources the world over: not only can roads be built to areas to be exploited but the hardest of natural materials can the more readily be broken, minced and generally got into shape for human use. Thus we have the modern food system, with fossil fuel energy needed at all stages from the manufacture of fertilisers through to the refrigerator cabinets in supermarkets. Further, those societies which do not possess these kinds of systems certainly aspire to them. There is however still a place for the sun and its gifts: much food still is based on photosynthesis, although the fossil input may eventually be the higher by the time the table is reached. Resource uses such as modern sea-and-sand tourism may also be said quite truthfully to be based on the sun, just as many other recreations now require a great deal of gear derived from industrial sources: consider winter sports for example (Plate 2.3).

Lastly in this grouping, there are resource-using systems derived mostly from fossil fuels and nuclear-generated electricity. The implements of these systems are the computer and other telecommunications, and urban transport networks, and the sites of resource use are the centres of the 'post-industrial' cities like New York, Tokyo and Hong Kong. Further developments include the hi-tech 'futuristic' cities which are planned from time to time: offshore complexes are very often reported to be on the drawing board in Japan. Pleasure again is involved. Apart from the dependence of say a modern squash court on a high-energy industrial base (for its materials and its lighting), the kind of tourism based on cities does not depend much upon contemporary solar energy. Indeed after a while most visitors to Rome or New York are very pleased to escape solar radiation and exchange it for electrically-powered air conditioning in the nearest café.

So there is a mosaic of energy intensities over the surface of the earth. The sun's contribution at the actual solid-gaseous interface is not even, and humans add to those flows in spatially uneven ways as well. We could almost say that human geography is founded upon the study of the spatial pattern of these combined energy flows.

Sources of energy for human societies

This major section looks at the main energy sources which can be tapped and transformed using today's technology (Table 2.7). We shall deal with the resource distribution, the uses of the energy, and the wastes created, though the major environmental links of the whole process form the basis of a later section. A basic distinction is made between renewable and non-renewable sources. The former are, of course, largely dependent upon the sun.

Plate 2.3 Not all energy is used for serious economic purposes. Some of it in the Western world goes on pleasure as seen here at Mount Hutt, New Zealand, in June 1990 where it is used to produce snow to form better ski-slopes.

Renewable sources of energy

About half the world's population relies on non-commercial fuelwood as its sole energy source other than food and for a larger proportion, biomass is their major source of energy. In fact, they consume most of the annual combustion of 12.5×10^{15} kcal, a sum which is ca 20% of the gross fossil fuel consumption in the world. In the developing countries of Sub-Saharan Africa, the contribution of fuelwood is often about 80 per cent of the nation's overall energy consumption and as Table 2.8 shows, the net contribution of biomass as a whole may be very large even in countries with less than a total dependence on this material. It is symptomatic of the scale of the use of fuelwood that the largest producers are India, China, Indonesia, USA and Nigeria. In the developing countries, fuelwood is needed for all cooking and space heating and is usually burned in simple stoves or open fireplaces of relatively low efficiency, or it may be converted first to charcoal, which is lighter and can be transported more easily, especially to towns. Combustion produces smoke that may be noticeable in towns, but elsewhere is only a minor problem; the ash is often valued as a fertiliser. Supplies of fuelwood are threatened in many areas where it is important, since demand is outstripping sustained yield levels. Thus some 100 million people are suffering an acute scarcity of fuelwood and a further 1.2×10^9 are living beyond sustainable levels of harvest: fuelwood here becomes a non-renewable resource.

Table 2.7 Sources of Energy for Human Economies

Type	Basic Materials	Distribution	Energy/ Mass	Initial product to economy	Scale	Non-heat wastes (recyclable, extractable)
RENEWABLE						
Organic	Wood	Ubiquitous exc v. dry or v. cold		Heat	Local, early	ash, smoke
	Other biomass	Products of		Heat	industrial	
	e.g. dung, leaves, straw	agricultural	Low	Motor fuel	Local, pre- industrial	ash, smoke
	Ethanol	economy Tropics			Regional	water, CO_2
Inorganic		Ubiquitous but		Heat, electric	Local/domestic	
Solar	Sunlight	better when cloudless		Electricity	Regional	
			var-	Electricity	Regional	
Wind	Moving air	High windspeeds best	iable	Electricity	Industrial	v. few: impact
				Electricity	Not known	visual, though
Tides	rhythmic water rise/fall			Heat, electric	Regional, local	
Hydro	falling water on land	Mountains especially				
O.T.C.	moving seawater	Tropical seas				steam and water
Geothermal	hot water 'hot rocks'	Seismic zones, 'hot rocks' at depth				with toxics eg sulphur
NON—RENEWABLE (FOSSIL)						CO, CO_2 oxides of sulphur,
Coal	Geological-age hydrocarbon	Many deposits on land and		Heat from combustion;		nitrogen, PAN, many other
Oil	Geological-age hydrocarbon	undersea. Worldwide but	High	explosive gases		gaseous and particulate
Natural gas	Geological-age hydrocarbon	discontinuous		(internal combustion engine)	Industrial	substances. eg. lead Radioactive particles. Refinery by-products.
NUCLEAR						
Fission-burner	Uranium-235	Uranium ore: is < 1% of geological ore	very high			Uranium ore wastes
Fission-breeder	Uranium-235	Process creates Pu-240 in larger quantities than it consumes.		Heat, then, electricity via turbine	Industrial Not known. Only of interest if very large	*Radioactivity* & de-contaminated materials Decommis- sioned power plants
Fusion	hydrogen-3 (Deuterium)	Sea-water: constantly renewed				Radioactivity

Biomass energy on a large scale has also become more popular where fast-growing crops can be converted to alcohol: sugar cane is the most popular of these especially since the decline in the popularity of sweet foods in the HIEs. Brazil mixes the alcohol with petroleum as a vehicle fuel ('gasohol') and reduced the share of oil in the national energy mix from 34 to 20% in the years 1975–84; in Hawaii in 1985, the island of Hawaii itself got 33% of its commercial energy as alcohol from sugar cane. In the USA, fuelwood has become popular as a domestic heating source, via high-efficiency closed stoves, but it has also been estimated that 25% of the wood entering the timber industry could be converted to energy: i.e. by revaluing the wastes. Industrial plants with an output of 20–50 MW are relatively common in wood-using industries. This attitude could well be extended outside the USA and some nations could generate between 12 and 50% of their total energy requirements from timber industry residues; e.g. Austria 14%, Canada 17% and Turkey

Table 2.8　Biomass energy in developing countries, 1978

Country	% of energy from biomass
AVERAGE LDOs	43
Bangladesh	71
India	53
Ethiopia	95
Niger	86
Tanzania	93
China	29
Brazil	34
Mexico	9
Nepal	98

Source: Adapted from WRI / IIED World resources 1987, NY: Basic Books 1987, p. 104.

42%. Likewise, waste biomass products like straw, coconut shells, peanut husks and cotton stalks can also be used as energy sources in many appropriate places, as can human excreta after fermentation in a tank from which the resultant methane can be drawn. The residue can still be used as a fertiliser, and the process points to the general desirability of using dung as a fertiliser rather than as a fuel, unless no crops are grown. Actually planting for example pines and eucalyptus as sources of both domestic wood and fuel for local industries has been undertaken in South Korea, Chile (over 750,000 ha) and Zambia (*ca* 45,000 ha), for example.

Although there is much scope for the development of existing ideas in the biomass field, it is necessary to emphasise the immediate problem of fuelwood supply in the LDCs. Shortages here affect so many facets of people's lives (especially of the women) that this must be reckoned to be one of the world's major resource supply problems. This is especially so in terms of the growth rates of demand and harvest (Table 2.9) which do not look sustainable in e.g. Africa and the drier parts of Latin America.

Biomass is mostly 'low-tech' in the sense that it is gathered and used without any great degree of machine-led complexity. At the other end of the scale of renewable energy

Table 2.9　Fuelwood and charcoal production, an average 1984–86

Region	$\times\ 10^6 m^{-3}$	% change 1974–86
WORLD	1,646,051	28
Africa	384,694	35
N & C America	154,387	169
S. America	216,961	26
Asia	739,144	17
Europe	56,508	9
USSR	85,567	3
Oceania	8,800	33

Source: WRI / IIED World resources 1988–89, NY: Basic Books, 1989, pp 288–89.

Table 2.10 Renewable energies: some capacities

Hydropower	World capacity 1985: 555 GW, of which 27% in N & C America
	Guri Dam, Venezuela; 10 GW Three Gorges, China (planned) 13 GW
Oceans	Wave power, Norway: 0.85 MW per plant Tidal power, France: 240 MW
Wind	California, 1987: 1.5 GW
Solar power	One Photo-voltaic plant, California, 1980's: 6.3 MW
	World solar installations in one year (1985): 25 MW
Geothermal	World 1985: 4700 MW electricity 1200 MW heating

Compiled from various sources

sources are those which need rather more technological intervention if they are to be harnessed (Table 2.10). This is especially so of hydro-power, a way of generating electricity which delivers about 25% of the world's electricity supplies. In seven nations in the 1980s, it represented between 90 and 100% of their supplies, and was over 50% in 45 countries. The advantages of this method have been the availability of finance for the installations which, although costing twice as much as coal, are 75% of nuclear, the longevity of the plant (twice that of coal), and the zero fuel costs. Thus in the LDCs over 2/5 of energy comes from hydropower, since energy-intensive industries can often be persuaded to move to relatively remote locations where large schemes (12 000 MW outputs are planned for multi-dam schemes in Brazil/Paraguay, for example) are possible. The social and environmental impact of such projects is immense and in China, for example, small installations of perhaps 80 KW are popular. Costs are perhaps $2000–3000 per kilowatt compared with $1500 per kw for a large dam but the intrusions are less and the control exercised locally. Hence, the potential for small-scale rural development is enormous. In Pakistan, a 10 KW plant will power 100 families; in the USA it would supply only two.

Hydropower also has a high remaining potential. Even if estimates like those of the World Energy Conference in 1980 (a quadrupling of capacity 1974–2020) and the World Bank (a tripling in the LDCs 1980–1995) come about, it seems that only 2/3 of the world's potential for hydropower will have been used.

Hydropower concepts can be transferred to the sea in the case of inlets with a large tidal range, where both flood and ebb tides can generate power: plants exist at La Rance in Brittany (France) and in the Soviet Union and feasibility studies have been undertaken for sites such as the Bristol Channel on the UK and the Bay of Fundy in eastern North America. The generation of power from waves, using various devices to translate the energy in coastal waves into electricity is also being pioneered on a number of high-energy coasts: it may have its greatest potential on small islands. Coastlands are also favoured sites (though not the only ones) for using wind energy to drive groups of modern windmills ('windfarms') to power turbines. The land can have other uses as well: slalom skiing is a rather jolly one, cattle grazing rather more prosaic. The great prize, however, is the ability to concentrate direct sunlight so as either to heat up water for use domestically (for hot water and heating, that is) or to generate electricity via a turbine or a photochemical reaction (Plate 2.4). Passive solar panels are now widespread, though rarely economic in cloudy climates; solar thermal plants are quite widespread in sunny climates where the concentration of the sunlight to heat up water to power a turbine can be achieved on most days and electricity can be produced for $4000–5000 per KW. In sunny regions, too, solar ponds can be used to heat water: power generation from these is possible but less frequent. These use the layering effect of salty water in ponds under strong sunlight and hot water can constantly be drawn off the bottom of the pond. They are cheap if land is cheap and so are sometimes found in desert areas where the ratio of 2.5–3.5 ha of land per MW is not significant, though the water supply may be a more difficult problem.

The earth itself is a potential source of heat, as the inhabitants of Iceland and Japan have known for centuries. Areas of volcanicity and earthquake hazard may provide some compensation in the form of underground hot water at high pressure that will drive turbines or at the least heat greenhouses and hot baths. More recent attention has focused on 'hot rocks': areas of volcanic rocks, usually granite, at considerable depths. Water piped down into contact with them can be brought up hot: some public buildings in Paris are heated this way. It is however, difficult to see how the cost structure will improve unless electricity can be generated as well. Human-produced 'earth' can also yield energy since landfill rubbish tips can be sealed and left to ferment anaerobically, producing methane that is drawn off and utilised. Solid wastes can also be incinerated to yield hot water and electricity, though relatively few cities as yet have ventured into generating their own power from their own wastes of all kinds. We might suspect that economics are not so much the barrier as a diffused feeling of distaste at reading by the light of one's own wastes, so to speak. This might not apply to materials like newspaper (4043 MJ/kg, cf coal at 4042 MJ/kg), waxed cartons at 5590 MJ/kg or polystyrene at 7525 MJ/kg.

Collecting energy from the sea is in general difficult but seawater does contain vast amounts of the hydrogen isotope deuterium (^{3}H) which might be the raw material (virtually limitless and in fact constantly renewed in runoff from the land) of energy from nuclear fusion. This creates huge amounts of heat and is currently only available in the form of thermonuclear weapons. If controllable fusion could be achieved then the resultant power might be truly renewable. Since the physics, chemistry and engineering are still at any early stage of development, it is treated under energy futures in a section below.

Non-renewable energy sources

This category comprises firstly the mineral hydrocarbon fuels of geological origin, the fossil fuels (Fig 2.2), and secondly the generation of electricity from uranium by fission, known popularly as nuclear power. Both these usages can stand a little qualification: the first category sometimes includes peat which is sub-fossil in type and age; and the second is only one way of generating energy from atomic nuclei but it dominates the commercial scene and so is identified with it.

No account of the hydrocarbons can omit coal. This was the founding fuel of the industrial revolution and has been responsible for the location of many industrial areas and cities throughout the world. Coal is found in lignites (or brown coals) mostly of Tertiary origin, through to the hard (bituminous or anthracite) coals largely of Carboniferous origin. The material is nearly all C, H, O, and N with variable amounts of sulphur and

Plate 2.4 Direct use of solar energy is a technology under active development in e.g., the USA. The sun is reflected from the mirrors to a central tower where high temperatures boil water. The technology is not without environmental impact in terms of land consumption. In these desert conditions it is not in severe competition for space, however.

is extracted either from open pits or by shaft mining laterally into hillsides or vertically to deeper strata. The largest reserves (Fig) are in Asia (especially Siberia), North America and Europe and on a world basis both hard coals and lignites have a long future at current rates of depletion, assuming that all recoverable deposits have been identified and that there are no enormous leaps in the technology of extraction. Although a bulky and relatively low-cost material, coal is transported over large distances: from Alberta to Japan and Colombia to Denmark, for instance. Hence there are big trade flows of coal just as there are for the more mobile hydrocarbons.

Oil, natural gas and natural gas liquids are the major remaining resource though there are deposits of tar sands and oil shales which attract some rather limited attention. These are often found together and often nowhere near the older established industrial centres,

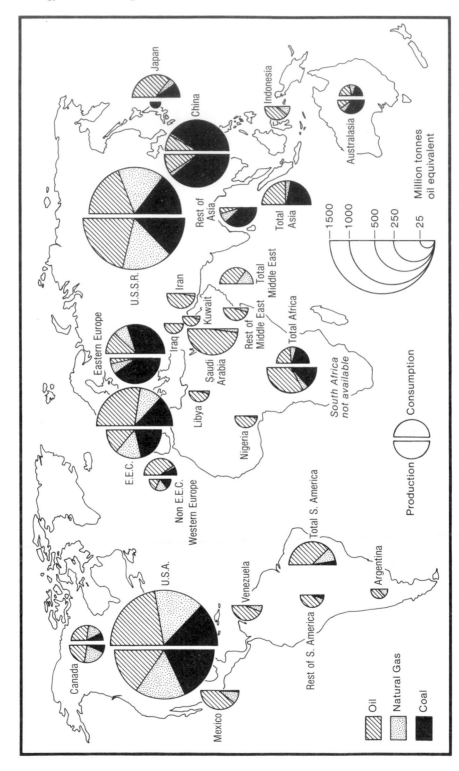

Fig. 2.2   World production and consumption patterns of oil, natural gas and coal, 1981, in million tonnes of oil equivalent. J. A. Rees, *Natural resources: allocation, economics and policy.* London: Methuen, 1985.

though newer complexes have been built around them, as on the Gulf Coast of the USA. As with coal, the material contains variable amounts of sulphur, mostly in oils. Gases such as methane are often not economic to use at the point of extraction and so are flared off. Oil is very easy to transport in bulk by tanker or pipeline but for maximum value has to be refined into various component chemicals before use. Several of these components are not used for energy but provide the stock materials for e.g. plastics and some pharmaceuticals. By contrast, natural gas is used as it comes, like much coal. Ease of transport means that there is an international trade which in value exceeds that in all other commodities combined.

The price of oil and natural gas is a key factor of the world economy. Petroleum products are the chief fuels of the western nations and oil exploration and recovery depends on the price as does the development of methods of improved recovery rates. On the other hand, lower prices on the world market are not always translated into for example lower kerosene prices in LDC markets, so that the relative attraction of kerosene and wood as basic fuels is not necessarily changed. The current rates of extraction and consumption (Table 2.11) (which are rising on a world scale by between 2 and 4% per year) would give life-times of 31 years for the oil resource and 58 years for natural gas but these would probably be expanded if the price rose since not all recoverable deposits have been discovered in an absolute sense and in any case the notion of what can be recovered is partly a function of price. Most forecasts of the price and the longevity of the resource have in general been wrong. There is, obviously, a political component in the price of oil. Producer countries form cartels to keep up the price in order to enhance their own incomes and consumer nations have a strong interest in the politics of the producers. These latter should ideally be stable and well-disposed to the customers; it is not unknown for outside interference in a country's affairs to be motivated by the perceived need to secure oil supplies.

The use of the fossil fuels demonstrated their ubiquity in industrial economies, being essential components of industry, transport, commerce residential developments, and agriculture. On a world basis, for example, 4% of commercial energy went into agriculture, and of this, 44% went into fertilisers, 2% into irrigation and 2% into biocides. As the latter, 160 kg/ha of oil equivalent ended up on the fields of the developed world, and 200 kg/ha oil equivalent on the agriculture of North Africa and the Middle East. A link to the energetics of modern food production (see pp 75–78) is therefore immediately apparent.

The next 100 years will undoubtedly see a move from the central role of oil in industrial economies, though it is by no means clear what will replace it since nuclear power exhibits a number of economic and environmental problems and the renewables do not match oil

Table 2.11 Global primary energy consumption: rates and shares

Type	Share in 1988 (%)	% change over 1987	% change over 1979
Oil	38	+ 3.1	− 3.3
Coal	30	+ 4.7	+ 23.4
Nat gas	20	+ 3.7	+ 26.7
Hydro	7	+ 0.5	+ 24.5
Nuclear	5	+ 8.0	+ 181.4

Source: E. S. Tucker, Growth rate accelerates, *Petroleum Economist* LVI(8), 1989, 243–45.

in terms of the same quantities of energy at the same kinds of price. There is also the complicating factor that, as oil in particular gets scarcer, it may be valued more for its material content than its energy. In these transitions, the relative roles of governments and the big transnational companies will be an interesting piece of politics; then throw in the interactions with global warming scenarios, the unpredictable aggressions of nations in oil-rich areas, unforeseen breakthroughs in the technology of alternatives, either the collapse or rapid resurgence of nuclear power, and the potential for energy conservation. All of which add up to a range of futures for fossil fuels being a more sensible approach than any simple extrapolation of present trends.

The current generation of nuclear power generators uses uranium-235 as its fuel. This isotope is only 0.7% of all naturally occuring uranium and indeed uranium ore is like any other mineral, finite in supply, though it is not normally thought of as limiting in the nuclear fuel cycle. The fission of atomic nuclei to generate heat and hence electricity (it does not deliver energy in any other form) is practised mostly in developed countries, with a few outliers (Table 2.12). Thus France and Belgium have the highest proportion of their energy budgets filled from nuclear sources and the USA the highest absolute use. (When reading data for these proportions always check to see whether the nuclear figure is a proportion of total commercial energy use or total electricity use.) Worldwide, the growth of nuclear installations (Plate 2.5) rose 99% between 1980 and 1985, but there is now a slowing of the rate of increase. This is due partly to public loss of confidence in the technology and its operators after some of the accidents, and partly because the promise of cheap electricity has rarely been fulfilled. Even in France, the citadel where 66% of electricity is generated by this method, EDF (Electricité de France) had a debt of F32 bn, which is larger than some LDCs.

So there are several interactions here: the price of uranium, the, as yet unknown, costs of decommisioning plants after their lifetime, the public attitude after accidents, and the fears over clusters of leukemia and other cancers provide an uncertain set of possible futures for fission reactors. The next step, technologically, is to breeder reactors which can use the uranium-238 which comprises 99.3% of the naturally occuring mineral. The economics of this are not clear and the process produces a lot of plutonium, so that no advanced nation is currently working very hard at this generation of nuclear plants; instead they would like if possible to miss it out and go straight on to fusion.

Table 2.12 Civilian nuclear power, 1980s: examples

	No of reactors	% of electrical power
WORLD	417	17
France	45	70
Belgium	5	65
USA	111	17
USSR	90	11
Japan	31	23
Bulgaria	4	32

Compiled from various sources

Energy consumption

Three initial points need to be made: first that consumption does not necessarily coincide spatially with production, so that energy has to be expended in actually transporting energy. Second, that some transformation of state may take place, as from chemical energy in coal to electricity, so that the price may be affected by the complexity of the transformation process and its efficiency. Third, consumption implies that the energy is changed into a form where it cannot do any work nor produce useful heat, i.e. it is radiated to space as low-grade heat; unlike materials it cannot be recycled.

The global consumption patterns of energy (Fig 2.3) are mostly compiled from data on commercial energy and thus are dominated by those nations which have major industries. The two basic yearly statistics (for 1986) are a global consumption of 273,201 PJ, which is an increase of 45% over 1970, and a per capita average of 56 GJ, which is 9% up on 1970. As we would expect, there are some considerable variations in consumption levels: at 58,422 PJ/yr the USA consumes more than the whole of Africa (16,941 PJ), with a per capita access to 278 GJ, compared with Africa at 12 GJ. Nepal scarcely features at all (2 PJ and 1 GJ/cap) because of the relative unimportance of commercial energy. The per head consumption records are held by producers such as Qatar (564 GJ) and Bahrain (429 GJ) which dwarf the consumers of Western Europe for whom the UK at 157 GJ and Portugal at 39 GJ may stand proxy. Recent global trends in consumption have been characterised by rises of between 2 and 4% per annum, e.g. 3.7% in 1988.

Non-commercial sources are not easily quantified on a national scale but, as was stated in the section on fuelwood, several countries in the South depend largely on recent organic materials: India, Brazil, China and Indonesia lead in this category, using $100–200 \times 10^{12}$ m^{-3}/yr each, but there are 17 other nations who consume over 10×10^{12} m^{-3}/yr. In today's world these represent a stage reached by the current DCs before the full flowering of the industrial revolution. At that time and subsequently, wood was largely replaced by coal, which was partly substituted by oil, with the addition of hydropower at a more or less constant level in recent years, and also nuclear power, which at the moment seems likely to fall in importance in all except those places like France which are very heavily committed to it. The less developed world, by contrast, still needs large supplies of wood and charcoal, and adds oil (sometimes as kerosene) and hydropower as national circumstances and foreign exchange permit. In 1975, for example, the sale of one tonne of copper ore on world markets would buy 115 barrels of oil; in 1981, only 57 barrels could be purchased.

The clear connection between access to commercial energy and a higher material living standard (Fig 2.4) provokes the question of what level of energy consumption is necessary to provide the physical foundations and intellectual opportunities for a dignified life. Using the reasoning of Vaclav Smil, the figure of 70 GJ/cap/yr seems reasonable. This is about four times that of the average for the poorer nations, 20% higher than the global average but only one third of the average for the richer countries. The datum of 70 GJ represents the levels of for example France in 1900, Sweden in the late 1950s and Japan in the late 1960s: none of these would seem unacceptable to the poor nations of the world, especially if some of it came as the versatile electricity. (Though it is argued that in places like Nepal, rural electrification means people stay up later and burn more wood; nevertheless TV has been described as the best contraceptive yet.)

One way of achieving better access to energy for the poor is said to be a lower rate of consumption by the rich, so that we next need to know how the industrial nations actually put their energy to work. The first finding is of course that a great deal of it does no work at all: many uses, such as the conversion of coal and oil to electricity, are highly inefficient; liquid fuels in autos waste 75–80% of the initial energy. Data are usually given by sectors

Plate 2.5 Nuclear power generation and full reprocessing are at the technological pinnacle of commercial energy supply. This photograph of part of the Sellafield (England) reprocessing plant (in 1990) emphasises the linkages to the environment, most notably the covered pipeline in the foreground which carries radioactive wastes to the Irish Sea.

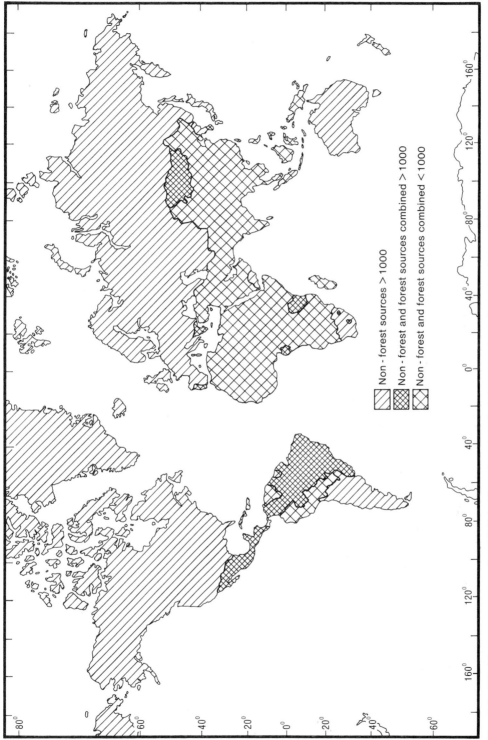

Fig. 2.3 World energy consumption 1970 in kilos of coal equivalent per capita, where the datum of 1000 separates different types of pattern. Earl, D. E., *Forest energy and economic development*. Oxford: Clarendon Press, 1975.

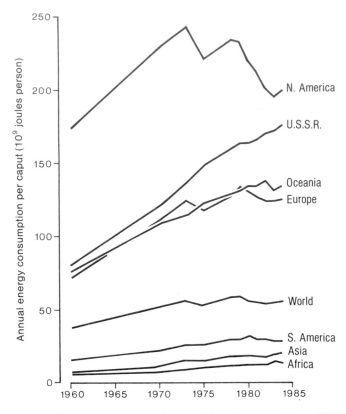

Fig. 2.4 Trends in per capita energy consumption: world and major regions, 1960–85. UNEP, *Environmental data report*. Oxford: Basil Blackwell, 1987.

like industry, transport, agriculture, commercial plus residential, and a miscellaneous category which includes for instance military use. Industry in the West typically uses 30–45% of available commercial energy, and transport 17–35%, with residential use at 20–35%. That these levels are not immutable is shown by the range of consumption in different versions of the same activity. In Italy, the production of a tonne of steel takes 17.6 GJ of energy; in the UK and the USA *ca* 23.5, in the USSR 31.0 and in India 41.0 GJ. Going further, these inputs for steel *de novo* are all higher than the 10 GJ/t needed for steel from scrap. Transport demands vary with the technology used: in Western Europe in the 1970s rail and bus transport consumed 400 KJ/passenger-kilometre, a car pool 650, one person in an auto 1800 and air transport 3800. Similarly there are differences in domestic consumption according to the degree of insulation of housing and the efficiency of domestic appliances, for example. A house in Texas that consumed 14,113 KWh/yr of electricity was retrofitted with better insulation and more efficient appliances and its consumption dropped to 5261 KWh/yr.

One way of codifying these uses is a measure of energy intensity, i.e. how much energy has gone into producing goods and services. This can be calculated (not without difficulty) for individual goods and services and for the whole of GNP. Thus in the 1980s, each US dollar of GNP (at 1982 value) needed the consumption of 20 MJ in Western Europe, 30 MJ in the USA and Canada, and 12 GJ in Japan. These data suggest that energy use by the affluent has a big role on energy consumption: driving for pleasure, flying to holiday destinations, having air conditioning and electric heating are all big dissipators of energy; one estimate has suggested that 21 per cent of the entire US commercial energy budget is devoted to the private automobile in all its phases, including manufacture, insurance, road construction and the like. Small wonder then that so much of urban life is structured round the auto even to the point where in the 1950s a large proportion of the population was actually conceived in one of them.

The idea of embedded energy in goods and services gives us some clues as to where economic growth might be achieved without creating an immense extra demand for energy. We can use the measure of GJ per dollar of value and discover that the energy costs (in the early 1980s) of water supply and sewage were 72 GJ/$ and cars 59; toys and sports were also high, at 51 GJ. In the 40–49 GJ bracket were hotels and food, and between 20 and 39 a whole range of activities including business, education, clothing, clubs, retail trade and insurance. Calculations for medicine vary between 11 and 34 but the theatre, opera and spectator sports are lowest of all at 10 GJ/$. (Further examples are given in Table 2.13). All this seems to suggest that a post-industrial economy has no need to provoke the exponential rises in energy demand which characterised earlier phases of the history of those nations. In fact, taken over the period 1973–1985, the energy intensity of many national economies fell: Australia and Canada by −6%, for example, Japan by −31%, the UK by −20% and the USA by −23%.

Table 2.13 Energy embedded in some products and processes, U.S.A. 1970's

Energy cost per cost of product (%)	Canned peas	9.4
	Fresh potatoes	20.8
	Fresh apples	11.0
	Cheese	3.0
	Ice cream	4.6
Fertilizer production (MJ/kg)	Nitrogen from ammonia (average)	60.0
	Pesticides:	
	Paraquat	459
	Maneb	99
	Propachlor	290
Machinery manufacture (excluding energy in metals) (MJ/kg)	Tractor	28
	Combine	22
	Planter	17
Containers (MJ/$ 1974)	Wooden	34,339
	Paperboard	72,688
	Glass	86,530
	Metal case	91,767
	Cloth bags	47,828
Miscellaneous items MJ/kg 1974	Binder twine	79.2
	Black plastic sheets	15.8

Source: R. C. Fluck and C. D. Baird, *Agricultural energetics*. Westport, Conn. Avi Pub Co., 1980, 87–119

Hence, the provision of more energy for the world's poorer people is not impossible, but it seems unlikely politically that the richer nations would revert to their levels of 20 years ago. Thus the capturing or liberation of extra energy from nature for those nations is still perceived as a desirable end.

Energy and environment

Energy provision and use has a major technological dimension, especially in industrial societies, but part of the ramifications extend far beyond the actual resource flow itself and indeed form part of the social and political fabric of a particular society. The effects of energy development and use have occasioned socio-political conflict over air quality, water supply and issues of human health. No energy delivery and use process is so free of environmental risk that it brings only gains and no disbenefits, and future impacts are so uncertain that precise prediction of the environmental effects of continuing growth of energy consumption cannot be undertaken with any confidence.

There is, it must be initially established, an outer limit to the environmental relations of energy. All energy transformations create heat and in the case of the combustion of fossil fuels and the fission of atoms, add to the atmosphere heat which would otherwise have remained in latent form. Thus to a natural heat flux in the atmosphere is added the production of heat from human-induced energy production and consumption. At present we perceive this in only local forms such as urban heat islands and by measurements from dense urban-industrial agglomerations which show us that in winter the added heat flux may be an appreciable fraction of the solar flows. At present the total added energy is about 1/5000 of the total solar energy received by the land mass but an added energy growth of 5% p.a. would bring added energy levels to those of the sun at the earth's surface in 2000 years. The global temperature would be about 40°C. Clearly this is an absurdity and so humankind must recognise and adapt to global climatological limits long before then.

The fossil fuels

Of the industrial fuels, coal has the longest history. It is won either by deep shaft mining or surface workings (opencast or strip mines) and both have impacts. Miners underground, for example, are subject to a high accident rate and to chronic respiratory diseases. Land transformation may occur with both types of extraction but restoration after surface mining is possible, though expensive; underground workings frequently cause land subsidence. Getting coal obviously changes the land surface, even if restoration and reclamation take place. As well, other resources are required: 1–2 tonnes of water are consumed for every tonne of coal delivered as electricity. Water flow through mines can result in very acid drainage which is also high in suspended solids and concentrations of metal ions. Oil and gas are more recent developments (Plate 2.6) and environmental concern has focused especially upon the effect of oil spills on marine life (in March 1989 the *Exxon Valdez* spilled 11 million gallons of crude into Alaskan waters, polluting 2900 km of coastline) and of leakage from oil pipelines as well as the environmental impact of their construction. Refineries are emitters of large quantities of airborne effluents (e.g. a variety of gaseous oxides) as well as waterborne substances like grease, ammonia and phenols. Notice how coastal producing regions necessarily coincide with areas of high biological productivity (Fig 2.5).

When fossil fuels are burnt to provide energy, a number of waste products are formed

Plate 2.6 Fossil fuels are in the kind of demand which makes it profitable (1988) to extract them from under offshore areas like the North Sea, which are not always hospitable environments. Methane is flared off from this production platform and underwater gas pipes may change fishing patterns. Birds on migration are also attracted to the lights on the platform.

which are usually emitted to the atmosphere. The most important are carbon dioxide, oxides of sulphur and of nitrogen, carbon monoxide, organic compounds, trace metals and particulate matter like smoke and soot. The actual quantity of waste depends upon the fuel being used, the temperature of combustion and the technology in which it is employed, including any special devices (of which Fluidised Bed Combustion is gaining more ground: it minimises sulphur and nitrogen oxide emissions) designed to reduce the contamination of the air. For example, natural gas is usually very low in sulphur, whereas oil and coal contain 1–4% of it, so that a shift in industry and in homes from gas to coal would increase the quantity of sulphur emitted. Similarly, coal produces twice as much carbon dioxide per unit of electricity generated as natural gas, and 30 per cent more than oil.

Once into the atmosphere, the fate of these substances varies. Carbon monoxide for instance is oxidized to carbon dioxide; trace elements like the lead added to petrol to decrease 'knock' fall out at decreasing quantities away from the source, although some is taken into the upper atmosphere and deposited in polar regions for example; particulates particularly fall out in a plume downwind from their source. Some are subject to long-range transport in the atmosphere: sulphur, nitrogen oxides, sulphates and nitrates are notable examples. Translocation over distances of 1000 km and more nearly always means that any problems are international ones, with emitters and receivers liable to hold somewhat different viewpoints on the processes. The eventual fallout of sulphur dioxide and of sulphuric acid, and of nitrates and nitric acid is the major constituent of the phenomenon of 'acid deposition', which is perhaps better labelled 'acid precipitation'. Certain values are not in dispute: the increase in the sulphur content of the air in parts of eastern Canada, southern Scandinavia, central Europe and the western USSR; the rising acidity of the freshwaters of those regions, and increased incidence of dieback of trees in Scandinavia and central Europe, are all examples. The linkages involved are still subject

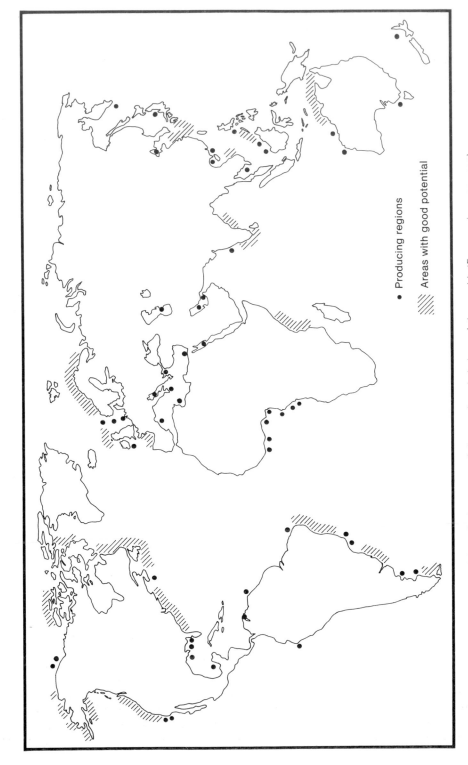

Fig. 2.5 Oil production and potential producing areas of the contiental shelves of the world. (From various sources).

to intensive research and no firm conclusions have yet emerged. Are forests in central Germany dying because of sulphur from British power stations or because of nitrogen compounds from German cars? Why is there acidification in Western Scotland, well away from most air masses that might carry sulphur compounds? What is the role of aluminium released to fresh water by more acid soils? These questions may in due course be answered: in the interim period many European nations have undertaken to reduce their emissions of sulphur by 30%.

At a more restricted spatial scale, there is the problem of the reactions between nitrogen oxides and hydrocarbons in the presence of sunlight, forming photochemical smog (PCS), which contains among other substances ozone and peroxyacetyl nitrate (PAN). It is characteristic of cities with high auto traffic densities and sunny climates. The 'type locality' is Los Angeles but many other cities, like Sydney, Mexico City, Lima, Tokyo and even northern cities like Edmonton and London have had regular episodes. One effect of PCS is to produce a brown haze which may reduce sunlight to 10 per cent of the incident value. Smog drifts out of the urban areas for distances of several hundred kilometres so that its effects on living organisms is not confined to human city and suburb dwellers, in whom it can exacerbate bronchial diseases, for example, as well as inflammation of mucous tissues. Where crops and forests are concerned, the constituents of PCS all damage the cells of leaves and hence impair photosynthesis. A weakened plant may then be less resistant to fungal or bacterial infection. When the Los Angeles basin experienced its worst periods of PCS, citrus fruits could not be grown within 50 km of the city and damage to plant life was reported over 3 million hectares of California.

Though not generally taken into account, the burning of coal produces radionuclides and one estimate in the 1980s suggested that coal was more dangerous than nuclear power in terms of total health impact by two or three times, if all forms of contamination (including but not confined to radioactivity) were considered.

The nuclear fuel cycle

Of all the current sources of energy, that generated by atomic fission has generated the most figurative heat. The hazard of radioactivity which is a particular feature of this technology (though not entirely confined to it: coal miners suffer from exposure to radioactivity) has created considerable public unease about this source of electricity and no proposal to build a new power station using nuclear fuel is without controversy, except in authoritarian societies.

At the heart of the discussion is the impact of ionising radiation upon humans and, secondarily, other components of the biosphere. There has always been radiation (the 'natural background') from rocks like granite, cosmic rays and gamma rays. To them has been added, without much stir, diagnostic X-rays. In the 1970s, the world population received an average whole-body dose of 1000 Sv/yr (Sv = microsieverts: the sievert is the S.I. unit of radiation dosage) from natural background and about 700 Sv/yr from man-made sources, of which < 50 Sv/yr was from nuclear power generation. In the UK the average dose from the nuclear power industry was 3 Sv/yr but exposed members of the population may have received 1000 Sv/yr from that source. One difficulty with all these data is that, although at high levels the dosage-response relationship is linear, at low dosages the relationship is not known and is difficult to distinguish from other variables, hence the reluctance of the UK authorities to admit the significance of high rates of childhood leukemia and other malignancies around military and civilian plant that handle uranium and plutonium products. It is possible that indirect linkages are present: one hypothesis is that male workers in higher radiation environments may father leukemia-prone children.

Radioactive wastes are generated in practically all areas of the nuclear industry's fuel cycle, from the mining of uranium through to the disposal of long-lived highly radioactive products. Some isotopes have a very short life (measured in hours or minutes), whereas others such as plutonium lose only half their radioactivity in 24,400 years. The 'front end' of the fission reactor fuel cycle creates large quantities of low-level wastes, e.g. in mining and milling, where the tailings and their runoff are contaminated. The concentration and fabrication of the uranium into fuel rods for the reactor produces relatively little radioactive waste, but the reactor itself and the spent fuel rods produce highly radioactive gases and liquids as well as being themselves contaminated. Thus the power station itself, when its life span of 25–35 years is over, has either to be dismantled and treated as spent fuel or be entombed in concrete. So from each stage, radioactive materials are led into the air or water according to carefully monitored international regulations, though these are sometimes exceeded because of operator error. In the end, a tonne of spent nuclear fuel (a year's production from a typical fission reactor) will yield 500 litres of hot highly radioactive liquid waste, which has to be stored above-ground pending final disposal. This latter is still a problem, though solidification in a vitrified material and burial in clays which lack ground water circulation seems the most likely option. Such materials will need to be isolated from the biosphere for about 0.25 million years, which is a remarkably optimistic commitment by any society. Most public concern focuses on the reactor itself. If the cooling systems fail then the reactor core containing the fuel rods may melt down, breaching the containment structures and releasing a plume of radioactive steam. One or two near-misses, including the highly publicised incidents at Three Mile Island in Pennsylvania in the 1970s and at Windscale in Cumbria in the 1950s, and at Chernobyl in 1986, have increased the public's wariness of nuclear power stations, and there is concern too over the transport of nuclear materials by rail, road and air, and the possibility of terrorism including the effects of a *kamikaze* impact by an aircraft. The probabilies of these incidents, especially of core meltdown, is estimated to be very low but as with aircraft crashes the actual casualties are likely to be very high, with genetic effects appearing a long time into the future. Thus many informed people as well as the public generally are unhappy about nuclear power and its attractiveness seems to have decreased during the late 1980s remaining strong only where pushed by central government as in France and Japan. Opposition to nuclear power is usually dismissed by the industry as 'emotional' but it is equally difficult to see what other adjective could be applied to their advocacy of it.

Overall, therefore, nuclear power possesses some of the environmental links of fossil fuel power generation. Land must be transformed, large quantities of cooling water for the turbines must be available, and much more heat per unit of electricity generated is given off. This hazard can be managed: indeed probably more forethought, care and skill goes into managing nuclear power generation than any other civilian enterprise. Even so, Murphy's law still seems to apply.

Hydropower

The large scheme is well known for its environmental impacts. The impoundment of the water may lubricate faults in tectonically unstable regions, causing small earthquakes, and in many climates substantial quantities of water are lost by evaporation from the reservoir surface. Downstream, the river regime may be altered in terms of flow (including the decided benefit of eliminating floods) and of temperature and silt content. If irrigation is a by-product then the increased availability of water may allow the breeding of malarial mosquitoes and the transmission of the larval stage of the fluke-borne disease schistosomiasis. At least seven other diseases are known to increase their incidence in humans in such developments. The construction of a dam and impoundment disrupts existing patterns of land use and settlement and in LDCs considerable social impact is also likely. On the other hand, the new lake may often create a very profitable fishery and in DCs the enhanced recreational facilities are popular.

Alternative energy resources

One of the principal reasons for developing the equilibrium or 'soft' energies is that they are thought to lack the heavy environmental impact of fossil and nuclear power systems. However these new technologies are not entirely benign: geothermal power for example may be the cause of land subsidence, may emit hydrogen sulphide into the air and a mixture of heavy metals and salts into the runoff. Waste water may have to be subjected to disposal by reinjection and sulphur may have to be precipitated out in order to reduce the effects on the local environment. Wind power systems if arrayed in any quantity are likely to produce considerable visual intrusion into coastal and upland areas of high landscape value and there are the possibilities of noise and interference with telecommunications. The potential environment effects of wave power collectors might include ecological changes in the calmer water shoreward of them, and visual intrusions. Tidal power generated on a large scale would produce ecological and social effects comparable with large dams inland: the quality of the water flowing into the dammed estuary would be a major consideration as would the source of building materials for the dam and the consequent changes in transport networks and associated urban-industrial developments. While no waste heat would be added to the atmosphere, this type of development would require an initial input of very large quantities of fossil fuels and be an enormous producer of change of every type.

Solar energy gathering by means of reflector systems constitutes an obvious change of land use and their industrial nature cannot be disguised. However, such installations are quiet and presumably the risks of intruders getting fried are acceptable. Photovaltaic systems have yet to receive much assessment of an environmental nature. Very hi-tech systems such as the idea of a satellite permanently beaming a concentrated ray of sunlight to an array of photovoltaic antennas might be acceptable until the beam of light missed the receiving station and withered the local crops.

Biomass energy is an obvious source of energy in HIEs as well as the poorer parts of the world: one estimate suggests that 11% of the consumption of the USA in 2000 AD could come from this source. But biomass energy cropping is subject to the same sort of environmental relationships as most other crops. If the throughput of water is high, for example, then local water-tables may be lowered and water not available for such consumers as livestock. If the cropping cycle is short then the soil may become depleted of nutrients and require chemical fertilisers, themselves consumers of fossil fuel based energy. Rates of soil erosion might increase as well. If plant material is fermented (e.g. to alcohol) then organic and mineral byproducts may result if they are not further anaerobically fermented to useful products such as feed additives or methane. But the largest potential difficulty may be land-use competition with food crops.

Overview

Just as all human-induced processes which yield resources have the capacity to change the natural components of the environment to a greater or lesser degree, so does energy usage. With energy however there can be truly global consequences: fine aerosols such as lead from car exhausts in the northern hemisphere find their way into polar ice-sheets, particulate matter may be carried up to the stratosphere and these reflect some solar radiation and carbon dioxide levels may have the capacity to change the entire global climate with all the consequences that could follow. No more convincing demonstration of the nature of energy use within the systems of the planet and human modifications of them can be found. This leads more or less directly to the benefits of energy conservation in DCs. Most areas of life here could be subject to vigorous measures to use less energy: buildings, transport, manufacturing and packaging are obvious examples. As a homely example, a 100-watt light bulb will over its life cause about 100 kg of CO_2 to be emitted into the atmosphere. A 25-watt compact flourescent bulb will cost less over its lifetime and use one eighth of the electricity, reducing the CO_2 output to about 12.5 kg over its lifetime.

Energy futures

It seems probable that the future of energy supplies is intimately bound up with that of other resource systems. Near-term choices for the developed nations are rarely unindirectional even where energy is a state monopoly, still less where private enterprise follows the pull of the market. One option in the face of the non-renewability of fossil fuels is an emphasis on nuclear fission as a source of electricity. France is committed to this path and the UK has strong advocates of it, though there is currently (1990) considerable uncertainty in view of the impending privatisation of a nationalised industry. Other nations are currently placing great emphasis on the conservation of energy and the using of it in the most efficient manner possible. They are not usually enthusiastic about nuclear power, preferring to develop more hydropower or alternatives instead, as in the case of for example Austria and Sweden. But penetration of alternatives is often slow: in the US with its great environmental and investment opportunities, only about 0.7 per cent of its commercial energy came from such sources in 1990, and that mostly from geothermal developments. For the LIEs, such choices would seem a great luxury. Most of them need more fuelwood and are trying to adopt both national and bottom-up development strategies that will produce more of this from a vibrant rural economy. The day of the great dam seems about over: most international agencies will not now finance them since the environmental and social costs are seen as being too high. Nuclear power is scarcely a possibility for most of the poorer countries, and so only incrementally can they hope to claw their way upwards in the consumption league, especially those well below the 70 GJ/yr level.

A mid-term prospect of great significance is that of nuclear fusion power. At present the fusion of light atoms of isotopes of hydrogen (the same reaction that occurs in the sun) can be produced but only in a largely uncontrollable fashion, known as a thermonuclear weapon. This fusion reaction produces immense amounts of heat and the appropriate hydrogen isotope, (deuterium) is found in sea-water where it is continually being created. Thus a virtually unlimited supply of energy appears to be promised by this fusion process, though lithium-6 might also be needed and that is not so abundant. The difficulty lies in the engineering since fusion only takes place at very high temperatures (1×10^8 to 1×10^9 °C), at which everything is a plasma in which atoms are stripped of their electrons. Containment must therefore use non-material forces such as magnetic fields or lasers. The complexity of running a commercial plant, if one can be developed, is such that fusion seems unlikely to be ever a cheap and ubiquitous source of energy, though large amounts of scientific expertise and money are still being put into projects in Western countries. But promised for about now some 40 years ago, it still seems that far off; there was a flurry of excitement in 1989 when two electrochemists claimed to have developed 'cold fusion' using heavy water and palladium in a kitchen-sink type of apparatus; however their findings of an energy surplus and the emission of neutrons proved difficult to replicate and the physicists have resumed their control of the development process.

In the face of medium-term pressures, one sensible strategy for many societies is to diversify. This provides a greater degree of security of supply but does little to ameliorate environmental impacts. In the early 1980s some 80 nations moved in that direction.

The long-term future of the energy resource must eventually pay some attention to the concept of entropy. In this context, this refers to the fact that high-grade energy (able to do work) is dissipated by life and by human structures. The creation of this more probable state of heat unable to do work is called the creation of entropy. Because low-entropy sources are limited (except for solar radiation), the rate at which we create entropy has social consequences: every sword forged means foregoing a ploughshare. Using this concept, it is possible to create a set of ethical precepts for minimising the creation of entropy which thus provide a framework for the greater sustainability of human life on earth.

Given all the uncertainties about the ecology, economics and politics of supply, the appropriateness of energy conservation seems incontrovertible. Added force to this movement is furnished by the predictions for the destabilisation of global climate fuelled largely by energy-use emissions. In fact there have been great gains in energy efficiency over the past 15 years. Between 1973 and 1985, the oil use of the OECD nations dropped by 15% while GDP rose by 21%; in the same period, Japan also achieved a per capita GDP increase of 21% but increased energy use by only 6%. Its steel industry, for example, cut the amount of energy required to make a tonne of steel by three-quarters between 1973 and 1986. The gains in domestic conservation and in buildings are well known, and many domestic appliances can achieve much lower rates of electricity consumption: an 80° reduction in refrigerator consumption is feasible using existing technology. A study in the USA showed that investing $12.5 bn in refrigerator efficiency would save $60 bn in avoiding the construction of 30 GW of new generating plant and (assuming the new power stations would have burned coal) save the emission of 160 million tonnes of carbon dioxide, about 4% of current US emissions. It is not impossible to imagine a scenario, therefore, in which global carbon emissions are held to present levels or even cut while allowing economic growth in the poor nations of the world.

A last word about energy futures must emphasise the poor track record of forecasters and prophets in this field. Many statements about the future, both pessimistic and optimistic, have been made this century and few have turned out to be true. Almost every new energy source ever, from steam to nuclear, has at some time been trumpeted as ideal, inexhaustible and about to introduce Utopia. So a certain dignified scepticism is always in order when talking about any time period beyond the next 20 years; what is not in doubt is the central role of energy availability and price in most of the resource-environment interactions of vital importance to human societies.

5

Renewable resources

The nature of renewable resources

In this chapter we discuss those resources which, if properly managed, can constantly renew themselves: food, forest products, water, and some living resources of the oceans. They are renewable for a number of reasons, the foremost of which is that several of them involve living matter. One of the characteristics of life is reproduction and so these resources will replicate themselves providing both the natural conditions and any alteration of them made by humans are right. These conditions usually involve the presence of pools of nutrients such as carbon and nitrogen which in natural ecosystems are cycled on a variety of spatial and temporal scales. Water is finite in amount but present in large quantities at several stages of the hydrological cycle which are accessible to human use and if properly managed is, again, self-renewing. In systems involving both water and living matter, the time period of natural cycles is important since it determines for example the growth rate of trees or the reproduction rate of animals. If crops are taken too quickly in terms of these cycles then the capacity of the systems to maintain a sustained yield is impaired.

In these resource systems, solar energy is critical since it is still a major and irreplaceable input into most food-producing systems; it is the basis of tree growth; it is the fundamental feature of oceanic ecosystems; and it drives the global climatic machine of which the hydrological cycle is an integral part. But human-directed commercial energy is increasingly involved in these systems: in providing fertilisers for crops, for instance, and in processing food for urban markets; in tree harvesting; in trawling the deep and remote oceans and preserving the catch; in pumping water uphill to irrigation projects or making dams to control floods. The proportion of the two different types of energy is of course variable through time and space. As far as organic matter is concerned, we can estimate the extent to which human societies have appropriated a renewable feature of the Earth's biosphere like Net Primary Productivity. By estimating the balance of natural ecosystems against those transformed by human activity, it can be seen that out of a global total NPP of 224.5 Pg/yr (1 Pg = 10^{15} grammes), 60.1 Pg are under human influence. Most of this effect falls on the land so that the proportion is more like 58.1 Pg out of 149.6 Pg, i.e. 39%. The remainder is left to sustain all other forms of life. So as a species we are clearly not living off the usufruct of natural systems: we are heavily involved in their control, often with a view to maximising their yield in the short term.

The food system

Humans will eat almost everything that they can get as far as their gullets. For small children this seems especially true, but it is adults which have encompassed the whole range from geophagy to cannibalism. Yet in spite of the fundamental importance of our nutrition, and the pleasure that most people get from eating and drinking, consensus about the ideal diet is still lacking. Thus statistics about the malnourished and the undernourished have to be treated with caution, while not denying that such conditions exist. We all need energy-yielding carbohydrates, proteins, vitamins and minerals but in exactly what quantities is difficult to answer since there are so many variables of activity and individual tolerances, as well as obvious differences between those maintaining their body in adulthood and e.g. children and pregnant women. Boiled down, so to speak, it seems as if the average adult person in the world (a truly tremendous statistical abstraction) who gets between 2200 and 3000 kcal/day, with 1/3–2/5 of this as animal protein, is likely to be well nourished. That many are inadequately nourished is as clear as is the fact that some are overfed, and the two problems may be connected; nearly all nutrition has an intimate connection with environmental processes and so the supply of food can still be regarded as a facet of the resources-environment field in spite of the great increases in production in recent years.

Major sources of human food

Before the evolution and spread of agriculture, food was collected rather than produced. Wild species of plants and animals formed the basis of nutrition and indeed these cultures (descendants of which persist in highly modified form, e.g. in Amazonia, the high Arctic and central Australia) are collectively called 'hunter-gatherers'. The term is sometimes extended to include 'fishers', which serves to remind us that much of our modern fishing industry is pre-Neolithic, for in spite of all the technology the men go out and hunt the fish. Only fish farming and aquaculture have made it through to a form of food production.

Food production depended in the first place on the domestication of appropriate species of plants and animals and though refined this has not basically changed, albeit with the advent of industrialisation new avenues of production have opened up and old ones been widened. Throughout history, human societies have sought to evade the limits imposed by nature upon food production. These constraints have come mostly in climatic form, with slope and soils as secondary contributors. Temperature is difficult to modify except by various forms of shelter like windbreaks and glasshouses, but water availability can be expanded by irrigation, slopes can be terraced, and soils can be given fertilisers. Yet unlike hunting and gathering, there are some environments where food production is not feasible unless ridiculous measures are taken: we could grow asparagus on Ellesmere Island if we applied the same sort of mentality that made moon-landings seem so important.

Time, culture and environment have sorted food production into a number of systems. Shifting cultivation is one such, generally a relatively simple system involving the temporary clearance of wild vegetation (grassland or forest) and its replacement with cultivated plots. In the tropics, animals are not usually involved as part of the cropping cycle. More common in the world now is permanent agriculture, where land parcels are devoted to raising plants and animals. Animals are often an integral part of maintaining the fertility of the soils in these systems, especially where chemical fertilisers are little used. Pastoralism and ranching are systems which rely largely upon animals, whose products are traded for plant materials. Any one of these systems may be supplemented by horticulture, particularly of the backyard variety. Lastly, there is industrial production of food which uses by products of e.g. petroleum refining in connection with micro-organisms to pro-duce edible substances.

Globally, a large proportion of our total diet comes from cultivated grasses and other grains, and just over 10% from land animal products. Such a reckoning, of course, conceals large national and social variations of intake.

World food production

The overall success of the world food system in feeding the growth of world population cannot be denied, though there are commentators who are less optimistic about the future. The recent past, in particular, has been a period of rising output on a global scale: in the 20 years to 1989, root crop production increased by 0.8% p.a., cereals by 3%, milk, meat and fish by 2% and other foods by 2.5%. In absolute terms, this meant for instance an increase in world production of grain from 1.0×10^9 t in 1965 to 1.87×10^9 t in 1986. Within this trend there was some yearly variation, especially in the USA where effective governmental policies determine the quantity of land under grains in a given year. In the LDCs most of the growth rates have been higher than the DCs, though from a smaller base, with India and China being the most effective higher producers. African production was stagnant until 1985 because of droughts and regionally it has been the only continent where production per head has in fact declined.

Unlike water, for example, there is a global market in agricultural produce, with large cash crop and export trades. Several nations earn most of their foreign exchange this way: Burundi gets 93% of its exchange from coffee exports, Sudan 65% from cotton. Overall, Africa devotes about 13% of its cropland to exports and most LDCs have a slightly higher proportion. To characterise trade, let the 1950 baseline equal 0. Then the volume of trade doubled by 1962 and quadrupled by 1980 and the value increased at an even faster rate. The trade is to some extent dominated by North American exports of grain, which provide the world with its buffer against harvest losses elsewhere, with 132×10^6 t/yr being a typical export loading which accounts for 87% of world grain exports. All this has hap-pened in an economic context of a fall in world food prices by 25% in real terms between 1950 and 1980 and also in a context where economic protectionism is widespread. For one reason or another the share of the LDCs in world agricultural trade has fallen from 63% in the early 1960s to 48% in the early 1980s.

This production surge has also taken place when the quantity of cropland per head is falling. The world figure is a projected decline from 0.39 ha/cap over the 1971–75 period to 0.25 ha/cap by 2000. In African LDCs the equivalent numbers are 0.62–0.32 and for South East Asia 0.35–0.20. In the developing nations, the fall is largely due to population growth and land problems (e.g. soil erosion, salinification of irrigated land), and in DCs such factors as over-production and industrial land uses will reduce the figure. Of themselves, these data need not be alarming since productivity can clearly increase: at what price we shall see later.

The global state, therefore, appears to be one of increased productivity keeping pace with population growth and even of passing the threshold of minimal self-sufficiency on a global reckoning. This, too, in spite of the fact that about 40% of agricultural production is lost in harvesting, processing and usage. But a universal transition to food security and more varied diets has yet to be brought about so that there are regional problems of nutrition, where some people are getting too little and others far too much.

Problems and responses in regional nutrition

From time to time, famine occurs in developing countries (Plate 2.7). Though symptomatic of long-standing problems of food supply, its immediate cause is usually perceived as some

Plate 2.7 Failure of a food system. A refugee settlement for Ethiopians in eastern Sudan. Failure of food production in Ethiopia in the 1980s resulted from a complex of causes, of which climatic change was only one.

'external' agent such as drought or civil war, and international aid can be supplied. Of more general importance are those LDCs which exhibit chronic undernutrition or malnutrition and especially those where intake is less than 2500 kcal/cap/day. Some 82 nations fall into this category, most of them in Africa, along with the poor areas of south Asia and Latin America. Ethiopia, Kampuchea and the Maldives all fell below 1800 kcal in the mid-1980s. The UN estimates (and we must bear in mind the difficulties of definition discussed above) suggest that 730 million people are getting insufficient calories and that in total some 950 million people live off deficient diets: i.e. 20% of the world population. The pastoralist economies of the Sahel belt of Africa have suffered especially badly from desertification and famine. Poor grazing control allied to pipe wells and animal disease control have been exacerbated by very dry periods. The reasons for these difficulties are not simple to analyse. Droughts, floods and woodfuel shortages together with rapid rates of population growth and the incidence of debilitating diseases are all implicated in some places, but not universally. Then there are social and economic factors; for example export crops may get government subsidies and so crops for local consumption are subject to negative discrimination, which will be made worse if the labourers are landless and hence subject to the wishes of a landlord. (Landless rural people are in fact the highest-risk group for poor diets.) Pastoralists may be subject to aid and government influences designed to maximise crop output to towns rather than improve the nutrition of the people themselves. Family structure may also dispose towards uneven nutrition: adult males may well get higher shares of the calories as well as any meat that is available: one study in Lagos showed

Plate 2.8   Non-failure of a food system. Abundant fossil fuel provides a space with controlled heating, lighting and supply of processed goods. It works for pets as well.

that adult males ate twice the protein of children of 7–12 years of age. The efforts of MNCs in discouraging breast-feeding are less than helpful, too.

At the other extreme are the overfed of the DCs, where there is the potential for an excess intake of animal fats, with links to the diseases of affluence such as artherosclerosis and heart attacks. Apart from the sheer variety of foodstuffs, the main element in the western syndrome is the availability of meat: steak probably represents only a 4% efficiency of conversion of energy of various kinds and even milk, a good protein source, is only 20% efficient (Plate 2.8). Keeping production high has led to large surpluses in e.g. the EC where in the mid-80s about 75% of the budget went on agricultural subsidies. Thus the productivity of DC agriculture, heavily subsidised by money and fossil fuel energy, means that the areas with 25% of the world population produce just over one-half of the food.

In the face of such problems, many solutions have been sought. There are those who affirm the necessity of technical developments in food production (for more detail see below), and those who are convinced that the problem is basically economic and can only be tackled by increasing the demand for food in the towns of LIEs so that better-off urbanites 'pull' more food out of the rural sector. Slower rates of population growth are a social factor which may back up any other approach. Socio-political developments like land reform are seen by yet others as the key to the success of any other developments. On the one hand, the virtues of indigenous farming systems and 'bottom-up' development are extolled and on the other large-scale technology-based early warning systems for drought or other extremes, like the FAO's GIEWS (Global Information and Early Warning System) scheme are also sought. One normative view suggests that the DCs should consume less in order to make surpluses available to poorer countries.

Agricultural development is by far the most popular of all these approaches, since it does not overtly mean changing the patterns of political power in a nation (though it may do so as a side-effect) and may be perceived as an element of a much-desired 'modernisation'. The post-1950 years have seen many attempts at increasing production in LDCs, either by extending the cropped area or by intensifying production. Common to both has been the practice of irrigation. Agroforestry has extended the growing of food crops to wooded areas and is seen as a deterrent to further deforestation especially in the tropics. More intensive cropping has necessitated disease control in both plants and animals, controlled breeding and many new varieties of crop plants, intercropping, and sophisticated systems of water control.

In effect, agricultural development has meant an attempt to transfer the technology of DC agriculture to the Southern nations, using a battery of HYV cereals (especially rice, maize and wheat), irrigation, chemical fertilisers and biocides, and water control. In terms of absolute productivity, the so-called 'Green Revolution' has been a great success but it has not eradicated all the pockets of poor nutrition simply because its benefits have not been evenly shared. Hence the popularity of improving indigenous systems which may benefit the poorest rather than making the rich richer. Nonetheless, irrigation has underpinned a lot of rises in production: in India 1950–1983, cereal output rose from 55×10^6 t/yr to 140×10^6 t/yr, with over 50% coming from irrigated areas. In fact, Asia housed 56% of the world increase in irrigated area from 166×10^6 ha in 1970 to 213×10^6 ha in 1981, with 257×10^6 ha forecast for 1994. Overall, the world irrigated total in the late 1980s was about 250×10^6 ha, which is 17% of total cropland. In terms of area per capita and lending by major donors to LIEs, the trend is however downward.

In spite of these technological applications, HYV varieties of cereals only account for one-third of the LDC plantings of such crops. Thus another 120×10^6 ha whose productivity has not increased for 20 years or more are still in crop. The challenge here is to use biotechnology to improve yields without producing the same economic and social dislocations as the Green Revolution.

The next large-scale development will be the biotechnology-based 'revolution', with genetically-tailored crops, animals whose genetic composition is patented, and perhaps even non-leguminous plants able to fix nitrogen from the air. Development of crop varieties able to tolerate salty and brackish water is already well underway. The effect of this upon socio-economic and political patterns is not predictable anywhere yet but it is indicative that a battle for control over plant genetic resources, as to whether they should be freely available to LDCs or remain in DC hands, was joined in the 1980s.

Everywhere except on a farm itself, food has to be bought or bartered and so the crux of the problem is making food available at prices people can afford. Part of the solution is technical but clearly the socio-economic-political complex has a key role in structuring the paths of production and reward in LDCs. In global terms, though, there is clearly a certain amount of slack in a system which can afford to devote to animal feed so much basic grain production and which could be sold at subsidised prices as part of development programmes. Yet again, one school of thought says that aid is the great retarder of lasting locally-galvanised development. While the argument proceeds, many go hungry.

Food and energy

Since food is our source of metabolic energy, the minimal requirement of a food system is to produce enough to keep a nuclear family alive, the children growing, and eventually reproducing. But surpluses are always desirable: non-productive individuals can be supported for their other talents or the extra food can be traded for other materials or favours.

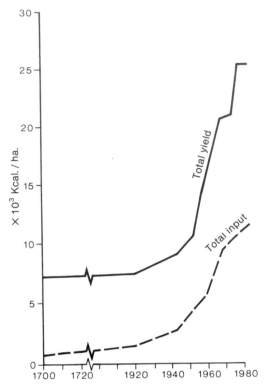

Fig. 2.6 A time-graph of the total inputs and yields in U.S. maize production, in thousands of kcal/ha. Output is related very largely to energy input in its various forms, both direct and embedded. (Compiled from various data).

This must have been the case in hunter-gatherer societies, although most of the energy in them must have been metabolic and not expended on 'cultural' possessions. Agriculture confers the ability to create greater surpluses: that of the Nile allowed the regimes of Ancient Egypt to export wheat to Rome, maintain cities with royal and priestly élites, and feed the men who constructed the pyramids. Indeed many great cultural achievements worldwide have been made possible by the surpluses generated by solar-based agriculture.

When the food system can be underlain with fossil fuel energy in all its forms, then its productivity per unit area/time or per person involved can be greatly increased. So an 'average' farm worker in the West can produce enough food to keep alive 32 others (Fig 2.6). Here we shall look at how energy finds its way into the food system and any lessons this has for economic and environmental management. A farm is the place where solar energy is taken into the system when fixed by plants as chemical energy. But to this is added more energy which derives from fossil fuels or in a few places from electricity generated in hydro or nuclear plants (Table 2.14). Some of this energy is applied to make the production of a crop possible (be it plant or animal) and yet more is applied after the crop leaves the farm on its route to the consumer; these two phases are called, respectively, 'upstream' and 'downstream'. In either case some of the energy is obvious: the diesel oil that fuels the tractor that ploughs the field, or that powers the tanker that takes away the milk, is an obvious case. But there is more to a thorough energy audit than those uses, for

Table 2.14 Energy use on US farms, 1977

Function	%
On farm:	
Machinery	18
Transport	15
Irrigation	12
Livestock	10
Cropdrying	5
Miscell.	3
Embedded:	
Fertilisers	31
Pesticides	5

Source: W. Lockeretz, Energy in U.S. agricultural production, in D. Knorr (ed). *Sustainable food systems*. Westport. Conn. Avi Pub. Co., 1983.

Table 2.15 Energy use in food production, USA 1977

Sector	Energy input $(10^{18}J)$	%
Production	1.6	13
Processing	3.8	31
Marketing & distribution	1.3	10
Home preparation	3.4	28
Eating out	2.2	18

Source: W. Lockeretz *op. cit.*

many products and processes (up and downstream) are only possible because of the availability of commercial energy and so contain an 'embedded energy' value. Fertilisers are a good example: energy has been expended in building the plant, in producing the materials, in delivering them to the farm by truck, in constructing a machine to spread them on the fields, and in supporting a salesman to convince the farmer of their value. Nitrogen fertiliser is in fact one of the most energy-intensive products used on a farm. Downstream, energy is used in building a drying plant as well as in running it and in making and running the family refrigerator which houses the sliced wrapped bread in whose slicing and wrapping further energy is embedded.

Using these concepts we can look at the whole food system as a consumer of energy (Table 2.15). In the USA in the 1970s, that nation used 17% of its commercial energy in its food system, with twice as much expended in the home as on the farm. In New Zealand, the food system consumed 30% of the primary energy and Australia 42%; the UK figure in 1970 was 23%. In each case about half of the energy is spent in the home. So a typical figure for the efficiency of the food system in a DC is perhaps 15% and if some of the commercial energy to produce the food has been imported then any country's food supplies contain an extra element of risk.

We may reasonably ask, 'does this matter?'. If richer nations want to devote some of their energy to having a varied diet with unlimited meat, then why not? Health reasons apart, eating less meat would be a form of energy conservation as would more efficient use of energy in the system as a whole, like using less packaging for example. This relates economically to the optimum depletion rate for non-renewable resources, and ecologically to the rate of production of carbon dioxide and its role in atmospheric warming. On top of this there has been a general ethical feeling that it was not right in the West to use so much energy on food when others were malnourished: this may have contributed to the undoubtedly greater energy efficiency in the food system since the data above were compiled but is minor compared with the greater debate about energy futures in general.

Agriculture and environment

This section will deal with agriculture as such, since the other parts of the food system are indistinguishable from other industries. We have to start with the everyday observation, confirmable by a visit to any farming area, that agriculture turns over the ground pretty thoroughly: the backyard, the animal pounds, the fields and their boundaries, wilder vegetation grazed by domestic stock, roadways and footpaths, are all evidence of changes to the pre-agricultural order, confirmed in place perhaps for some thousands of years.

If we look for the biggest global environmental changes caused by agriculture then two phenomena stand out. The first is the permanent removal of much natural or near-natural vegetation to make cropland. The outstanding biomes to have been affected are the temperate forests and the temperate grasslands, especially those of Europe, North America, the USSR, China and South America, but savannas, tropical forests and mountain vegetation also come within the ambit of affected places. The second is the commitment to permanence symbolised by terracing. The construction of terraces is a virtually world-wide phenemenon and requires much initial effort, whether this is purely human labour as the rice terraces of parts of Indonesia or machine-driven as with some of the recent re-terracings of the port wine district of the Upper Douro in Portugal.

If terracing is combined with irrigation, then change is even more complete. In a classic statement of the nature of wet-rice (*padi*) cultivation in Indonesia, the anthropologist Clifford Geertz compared *padi* fields to aquaria, and certainly pre-HYV *padi* yielded crops of fish and shrimps as well as rice, and the mud contained algae which were nitrogen-fixing. More modern irrigation, especially in semi-arid zones, changes the environment deliberately but many unwanted consequences are also produced. There are the human diseases discussed above but also ecological side-effects. The main one of these, where evaporation exceeds precipitation, is salinification of the soils and draws mineral salts upwards in the soil profile. Waterlogging of soils often follows, and the whole system is downgraded if the canals begin to silt under inefficient management of surrounding watersheds. Globally, as much land is being abandoned because of salinification as is now being brought under irrigation. The world area damaged is ca 60×106 ha, i.e. about 24% of all irrigated land; in India some 36% of irrigated land is classified as damaged by waterlogging and salinification. Even in a higher-technology nation like the USA, the equivalent figure reaches 27%. Some irrigated areas are supplied from underground aquifers and unless recharge rates are high, the water is 'mined' and is in essence a non-renewable resource; high rates of withdrawal, often found where irrigation and urban growth go together, may cause land subsidence.

Dryland farming in the DCs is not free from deleterious impact. Health questions may also be raised, centering round the various chemicals that may be applied to a crop, and also to the levels of nitrogen fertilisers that are used. Excess N_2 finds its way into groundwater and then into domestic supplies and has been implicated in nitrite form in infant blood disorders and adult cancers. On the ground, modern machinery is so heavy that it may compact soils and change their runoff, as well as agrarian, characteristics. Chemical fertilisers affect soil physics and chemistry since they contain no organic matter to contribute to the humus fraction of the soil. Biocides of various kinds (used in even greater quantities in the LDCs) may kill non-target species and build up in food chains with a variety of lethal and non-lethal effects on wild life, and both they and fertilisers reach the sea and produce the kind of ecological effects discussed in relevant sections later. The concentrated nature of much of today's animal production means that very large quantities of animal wastes (slurry in particular but also dead but inedible birds for example) have to be disposed of at a cost which does not negate the profitability of the farm enterprise.

The economics of the use of a great deal of machinery in western farming today has resulted in changes in field size, so that traditional landscapes with small fields have changed markedly. In places like England, this is very like a reversion to the large open fields of the medieval period, but cultural attachment to the enclosed, hedge-bound landscape in place by the mid-18th century is strong. Reactions to the loss of hedgerows and their flora and fauna, and to major species loss like the barn owl (no longer able to find nesting sites in modern buildings and finding fewer pollarded willows for the same purpose) have been strong, prompting government agencies to try and 'green' more farmers and agribusinesses.

Hence, all agriculture modifies the 'natural' environment but today's industrialised cultivation has very heavy impacts, often as thorough as those of industry itself or of towns and cities. Many minor changes have not been mentioned here and it is a useful exercise to go and sit for an hour in a rural area and make a list of all the visible and inferrable changes that have been wrought in the name of food production.

Futures of the food system

As with energy, it is unwise to think of 'a' future for food production. But any changes in direction are likely to be dominated by:

- A better food supply for the undernourished and the malnourished of the world: better in the sense of more secure and containing all the necessary elements, and preferably capable of being diversified.

- A lower environmental impact for farming. This is most noticeable in the DCs where it is driven by the dislike of landscape changes and the concern over chemical residues in food. In its most developed form, it is called 'organic farming', which would probably meet domestic demands in the HIEs with a lower cost of production but higher prices due to a lower total production. The latter, overall, is probably in the region of 90% of conventional production and may in fact be higher in the case of milk and bean yields.

Technological changes to deal with these concerns are various but we can confidently predict two types:

- To get to grips with the first category above, the resources of recent developments in biotechnology will be deployed. These will include, for example, crops which have been 'tailored' for particular circumstances. Resistance to particular pests or strong stems to combat high winds, the ability to tolerate salty soils, and for non-leguminous species to fix atmospheric nitrogen are all within either the ambit or the ambitions of genetic engineering now or in the near future. Reservations have been expressed over the release into the environment of genetically-manipulated organisms, since the mechanisms for monitoring and regulating such releases are by no means effective everywhere in the world. The history of exotic introductions (e.g. the starling into North America or the water hyacinth into tropical Africa) gives pause for thought.

- The conservation and battle for ownership of plant genetic resources to provide raw material for genetic engineering and to re-introduce 'traditional' varieties into farm patterns of HIEs, especially where organic farming is re-adopted to satisfy the demands articulated in the second category above.

None of this looks like a scenario for cheap food: as the Green Revolution showed, higher production does not necessarily mean that the poorest will benefit by better nourishment, or indeed in any other way. One crucial point will be the commercial energy needed to deliver the products of the genetic revolution to the farmers: if high quantities are needed then the social effects could mimic the experience with HYVs, which in very general terms was good for the already well-off. There is also the possibility that patented genotypes (which the 'new' varieties will be) will be expensive. So there is no guarantee that technological change will benefit those who are in most need of better feeding: it does not dissolve socio-political hierarchies and hegemonies. In fact, there is already food for all in the world if it were evenly distributed and even more if less grain were used for animal feed for richer countries. There is of course no copper-bottomed guarantee that such surpluses will continue. Cultivated areas are not likely to expand and technological fixes are as yet uncertain; a warming world may produce more droughts and diminished mid-latitude grain yields. Agriculture sits in a web of 'upstream' and 'downstream' environmental effects (Fig 2.7).

No surprise, then, that advocates of sustainable increases in food production are keen

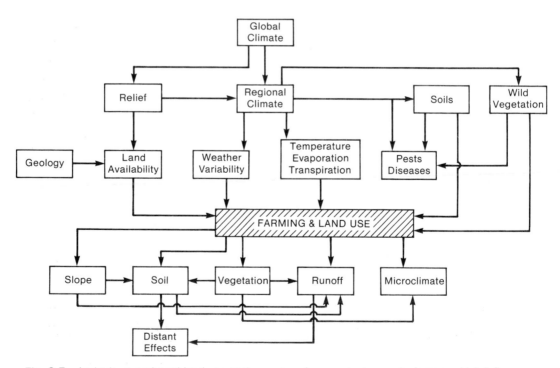

Fig. 2.7 Agriculture and rural land use at the centre of an ecosystem web: factors which influence them (above) and components in turn influenced by agriculture (below).

to re-assess the virtues of 'traditional' farming in the LDCs (Plate 2.9) as a way of tapping the knowledge of millenia and harnessing it to a minimum of 'appropriate technology' in order that a 'bottom-up' development strategy can bring benefits to all members of a local society and not just the powerful; socially a revaluation of women is often part of this approach. If the move towards less technological dependence in both HIEs and LIEs coincide and neither are swamped by the power of big business then food production could be the scene of the greatest revolution since the Neolithic.

But sustainable production in LIEs in the wake of the Green Revolution is a high priority. Development of this kind must stress social features at the expense of the purely technological. Thus the prioritisation of the farmers' needs, the security of rights and gains for the poor, and the importance of self-help, all feature in the more successful stories. Similar attitudes might not come amiss in some of the more heavily industrialised food and fibre production systems of the world as well, if, of course, production itself is not swamped literally by rising sea-levels or by rapid population growth.

Forest resources

Adaptable to almost all environmental conditions except the very cold, the tree provides not only a variety of useful products but has also featured heavily in the symbolism of human cultures: as a source of forbidden fruit as well as of life itself. In fact, the largest trees (of the genus *Sequoia* now found only in California) are the biggest living entities of

Plate 2.9 In the Dominican Republic, some communities have turned away form large scale agriculture to multi-crop husbandry with an emphasis on local needs and indigenous skills. This is 'bottom-up' or 'people-led' development.

both plant and animal kingdoms, the latter being headed by the blue whale. On an ecological time-scale, many trees are aggregated into forest ecosystems, where they are the dominant organisms which set the boundary conditions for many other parameters, such as understorey vegetation, soils, microclimate, and animal life. For humans to utilise the trees is therefore potentially to change nearly all flows of the ecosystem.

Forest and woodland resources and their uses

To equate trees with forests is to forget that they grow in a variety of other habitats as well. Open woodland (i.e. lacking a closed canopy of branches), small woodlots and coppices and even lone pines have their place in the spectrum of wood-producing places alongside the major areas of closed (i.e. having a more or less unbroken canopy) forest. Woodland may also have two major types of origin: they may have been progressively modified from a 'natural' condition in the last 10,000 years or they may be plantations, essentially analogous to an agricultural crop in being selected and managed intensively just like a field of corn (Plate 2.10).

The world today contains *ca* 2800 × 10⁶ ha of closed forests, which is 21% of the land area. The addition of 4500 × 10⁶ ha of open forest, and woodland in areas of shifting cultivation, brings the total to 34% of the land area, though this figure is prone to considerable error. But closed and open woodland together are approximately three times the world's cropland area. The tropics contain 43% of the closed forests, the temperate zones 57%; the developed nations are responsible for 90% of the coniferous woodlands, whereas the LDCs have 75% of the broadleaved forest (Table 2.16). About half of the closed forests are in fact found in four countries: Brazil, USSR, Canada and the USA.

The uses of woodlands and forests are manyfold (Table 2.17). The greatest single use is still the provision of wood fuel for domestic use in the poorer nations: in 1985 the harvest was estimated to be $17 \times 10^9 \, \mathrm{m}^{-3}$ for that purpose, whereas industrial wood harvested amounted to $1.5 \times 10^9 \, \mathrm{m}^{-3}$ (Table 2.18). Because of its high bulk and generally low value, less than 20% of the industrial harvest is traded internationally. In peasant

Plate 2.10 Large-scale husbandry of pine in Brazil (1988) is enhanced by modern nursery techniques. Reforestation becomes more like agribusiness farming in its environmental relationships.

Table 2.16 World forest lands (million ha), 1980s

	Closed forest	Other wooded areas	Total forest Wooded area	(%)
Temperate	1590	563	2153	(34)
N. America	459	275	734	(40)
Europe	145	35	181	(38)
U.S.S.R.	792	138	930	(42)
Others	194	115	309	(16)
Tropical	1202	1114	2346	(49)
Africa	217	652	869	(40)
Asia & Pacific	306	104	410	(43)
Latin America	679	388	1067	(64)

Source: WRI *World resources 1988–89*, NY; Basic Books 1988, p. 70

Table 2.17 Main productive uses of forests

As environments

Watershed protection	–	Controlled runoff Storage Soil nutrient maintenance
Atmospheric regulation	–	Absorption of solar heat in evapotranspiration Sequestration of CO_2
Erosion control	–	Shelter belts, dune fixation, rehabilitation of eroded terrain
Land bank	–	Soil nutrient and structure maintenance

Local Uses

Fuelwood charcoal	–	Cooking and heating
Pole-size wood	–	Housing and other buildings, fencing, furniture
Sawn timber	–	Construction, joinery
Weaving fibres	–	Baskets, nets, furnishing, rope, string
Sericulture, agriculture	–	Silk, honey, waxes
Special woods	–	Drugs, incense, carving
Food production	–	Grazing, hunting, gathering, shifting cultivation, fishing

Export industrial uses

Gums, resins etc.	–	Naval stores, tannin, resins, turpentine
Poles	–	Pitprops, transmission poles
Sawn timber	–	Lumber, furniture, joinery, construction
Veneer logs	–	Plywood, furniture
Pulpwood	–	Newsprint, papers and boards, containers, textiles
Residues	–	Particle and fibre boards
Food production	–	Ranching for cattle

Table 2.18 World wood production 1985 ($\times 10^9$ m^3)

	Total	Round wood removals Industrial	Fuelwood
WORLD	3165	1502	1663
Developed Nations	1405	1151	255
Developing Nations	1760	351	1408

Source: WRI *World resources 1988–89* NY Basic Books 1988, p. 74

economies, the uses of wood are for fuel, construction, fencing, tools and a myriad of other purposes; in industrial nations, construction and furniture are major uses but the predominant demand is for paper and paper products of all kinds. The weight of these demands have led to the point where deforestation is proceeding at 10–20 times the rate of reforestation: in the last 10,000 years perhaps 33% of the broadleaved forests have been removed by human hands, and 25% of the savanna woodlands and sub-tropical deciduous forests. Until quite recently only 6% of the tropical moist forests had been felled. Thus unless intensive production can be maintained from the current area, the global status of forests as a renewable resource is threatened.

Apart from their yields of wood, forests have other values for human societies. Unless they have no understorey and ground-layer vegetation, they can often be used to pasture domestic animals. They hold the soil and thus act as a protective cover for watersheds (a quality recognised by regulation in Tokugawa Japan, for example), and are reservoirs for wildlife. In terms of products other than timber, a survey of intact TMF in Peru showed that one hectare contained 275 species of tree, of which 26 per cent yielded materials with a market value in Iquitos: one valuation puts the forest's worth at $6820/ha, compared with $3184 for monocultural tree plantations nearby and $2960 for good quality Amazonian cattle pasture. All the qualities of the forest come together in the popularity of the habitat for recreation of various kinds separately. (Wildlife and recreation will be discussed (Plate 2.11) Finally, forests may have regional climatic roles: in the Amazon basin, for example, much solar heat is used in evaporating the moisture produced by evapotranspiration from the trees; without them the region would probably be much

Plate 2.11 Multiple use of forests. This coniferous forest in Scotland is clearly popular with children, affording them a safe place to ride their strengthened bikes. Other recreations will include nature trails, wildlife observations, picnics and scenic drives.

hotter. Globally, the forests are one of the sinks for carbon. Without their role in sequestering atmospheric carbon dioxide, for instance, the concentration of that gas would be rising even faster than at present, with worldwide climatic consequences. Forests, then, are an integral part of the human life-support system, as much as food and water.

Regional problems of overuse

Understandably enough, when forest loss has been discussed in recent years, attention has been mainly focused on the TMFs, the woodfuel problem and the effects of acid rain. But there are other problem areas, albeit in a world picture which is by no means unrelieved gloom, since many nations have effective and profitable (in all senses) forest programmes and policies.

One other region of difficulty has been the Himalaya. The linkage between deforestation of countries like Nepal and the flooding of the Ganges valley and delta has been put forward as one of the globe's major eco-crises. The ecology of the processes consists of a number of strands:

- The population growth of Nepal is of the order of 2.6–3.0 per cent per annum and may not yet have peaked: the total is now over 18 million. This represents a doubling time of *ca* 27 years and occurs in a population which is 90% rural and devoted to subsistence agriculture; further the people lack access to energy sources other than those of organic origin such as fuelwood, brush and dung.

- The needs of the population are said to deplete the forest cover, leading to 50% loss of cover in the period 1950–80, with a prediction of total deforestation of accessible areas by 2000 AD. This process has been accompanied by the construction of agricultural terraces on steeper slopes and both have led to increased soil erosion, higher incidence of landslides, increased runoff in the summer monsoon and thus lower water tables in the dry season. Rivers silt up and change their course unpredictably, and spread sand and gravel across fertile plain-land during floods. These latter ills extend as far as the deltas of the Ganges and the Bhramaputra.

- The woodfuel is scarcer either because land has been converted to terraces or because the remaining forests have been cleared. Animal dung is therefore used for fuel and deprived the terraces of their main source of fertiliser. Productivity is lower and so more terraces are cut on marginal and steeper land; their location and the weakened structure of soil lacking fertilisers lead to more landslides.

Hence it is argued that Nepal is exporting to India the one commodity that it can least afford to part with (i.e. topsoil), and in the form in which India can least afford to deal with, as silt that fills reservoirs, clogs turbines and irrigation schemes, and causes floods. Yet a detailed case is made by Ives and Messerli (1989) that the causal linkages between population growth, deforestation, loss of agricultural land, and downstream effects are simply not supported by rigorous and reliable data. Especially, the whole region is characterised by a varied geography which forces a great degree of uncertainty upon any broad sets of hypotheses which construct a general theory for the whole of Nepal or indeed the Himalaya *in toto*. The lesson we bring away from such a detailed examination is that resource-environment linkages have their local and regional flavour (which extends to the solutions to problems where they exist), which may not be susceptible to aggregation for larger areas. So accounts of the forest depletion of 'the tropics' or 'Amazonia' need always to be heard with a sceptical ear which wonders just how much of the region is in fact affected by the processes being described.

The fact that not only developing regions have forest resource management problems is demonstrated by the case of the USSR. A number of publications in the 1970s showed how that great nation was creating a non-renewable resource out of a vast forest reserve. The forest resource of the Soviet Union amounted to 738.2×10^6 ha in the 1960s, to which could be added burned and unreforested logged areas to make a total of 910×10^6 ha. These were classified into three main types according to the relations between protection and availability for logging, with the intensive-extraction category occupying by far the greatest proportion; further, recategorisation was not difficult. In spite of many laws governing the management of the forest resource, the particular economic and institutional features of the Soviet system resulted in poor harvesting processes, considerable waste of the harvest, and big losses to pests and fire. For example, the conifer resource diminished by 5% during the 1960s because of failures to obey the laws about reseeding; in the RFSR in the 1946–60 period only 15% of logged land was reseeded and for the whole nation the figure was 30–35%. Also, a great deal of timber goes to waste since it is easier to fulfil production quotas from accessible areas rather than take out mature and over-mature trees from further away; this is a feature shared with unregulated capitalism, of course. Other undesirable features of the management include a loss of one third of the harvest, often in the course of rafting down the rivers, 1 million ha/yr consumed by fire and over 18 million ha lost to HEP projects, i.e. drowned in reservoirs. Paper and pulp mills have contributed their share of toxic wastes to air and water-courses (Plate 2.12), including the pollution of Lake Baikal, though action has been taken over this latter.

Plate 2.12   A Finnish pulp and paper plant in 1988. Energy is consumed, which can be generated locally from wood wastes. Emissions in this case include chlorinated hydrocarbons and dioxins. These are highly toxic to life in the lake on the edge of which the plant is found.

The problem in the USSR, then, is not one of science, since the principles of sustained yield forestry are well known, but the implementation of that knowledge in the face of a bureaucracy concerned only to produce designated quantities with no regard for the external costs of the processes. It will be interesting to see whether political change in the USSR and Nepal bring about different attitudes to forest use and management.

Energy and forest products

Given that forests are, with coral reefs, the most effective accumulators of solar energy that nature has evolved, it is the more interesting to compare their flows with those of the human societies that exploit them. One example of this is in the tropical forests which are used by shifting cultivators with a relatively simple technology, albeit combined with a deep folk-knowledge of the ecology of the habitat. Shifting cultivation has often been seen as an efficient way of using the forest since after the plot is abandoned the recolonising forest restores to some extent the energy and nutrient flows. This has been acknowledged in uses of the forest such as plantation woodland and agroforestry, where the yield per unit area to humans may be lower but is more sustainable than practices which involve the clearing of trees entirely. This acknowledges the services of nature (in this case, the trees) in conserving and mobilising nutrients. Thus any equation involving the energetics of shifting cultivation should deduct the energy present in the trees since this is lost by burning them off. If, with C.F. Jordan (1987), we estimate the calorific value of the trees at 1720×10^6 kcal/ha (7224×10^9 J), then the E_r for shifting cultivation changes from 14:1 to 0.005:1, which is a net loss system in these terms. Nevertheless, at low population densities, sustainable use of the totality of the forest habitat has been achieved in most TMF regions. It is in imitation of such processes that modern agroforestry (Fig 2.8) has been advocated.

Much of the complex technology of the 20th century can be brought to bear on the forest resource. This involves the delivery of extrasomatic energy into both forest habitats and post-harvest processing: felling and on-site processing, access to stands of trees, debarking and pulping, transport of low-cost materials like paper, site management processes like seeding, spraying, and fertilising (from the air, for example), and ameliorative measures against soil loss, all involve the application of embedded energy in machines and materials, if not the actual use of commercial energy and forest wastes. The balance of E_r can thus tip towards the energy-negative, rather like agriculture. At any one time, however, the energy content of the trees is so high that it far outweighs the energy of the harvesting process, if not that of the manufacturing of paper and other products. However the energy balance in agriculture is calculated at least yearly, that of the forests only after decades, and the two are difficult to compare. The Kraft-process pulp mill can in fact be energy-independent if all the wood wastes are burned in the plant to produce work and heat, though where electricity is needed it is often cheaper to buy this in.

Forest environments and multiple use

Ever since humans have lived in forests, their use of them must have had a number of strands, involving hunting, gathering, fuel and shelter as well as contribution to legends and myths. Even though the main product is now wood, other uses may be important and with suitable management can be combined with a régime devoted primarily to timber extraction (Fig 2.9).

An obvious example is that of the grazing of domesticated beasts, which can crop the ground layer of grasses and herbs, or browse on either undershrub or the lower branches of trees. Their main disadvantage is to inhibit or even prevent entirely the regeneration of

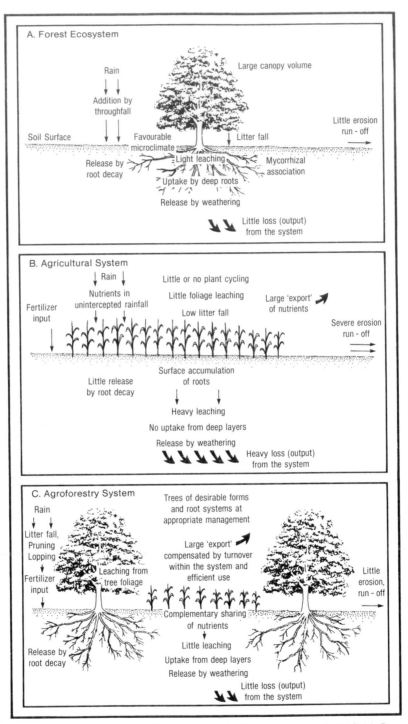

Fig. 2.8 A comparison of the nutrient flow in a forest (A) with an agricultural field (B) and agroforestry (C). M. Dover and L. M. Talbot, *To feed the earth*. Washington DC: World Resources Institute, 1987.

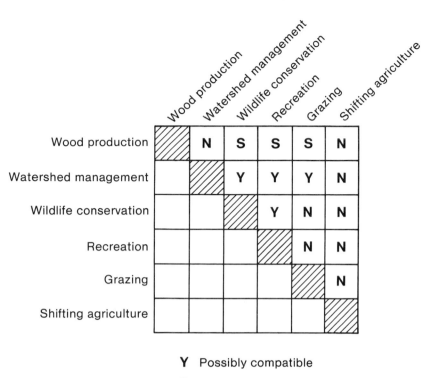

Y Possibly compatible

S Compatible if spatially separated

N Probably incompatible

Fig. 2.9 Compatibilities between different use of the forest as a resource.

tree species since they will probably eat the tender seedlings. Thus if natural regeneration is relied upon for forest renewal, careful management of exclusion zones is needed; more intensive management by replanting with nursery seedlings, for instance, would include fencing as a matter of course.

Other forest products besides wood include waxes and oils, and many drugs. Quite a number of major pharmaceuticals have their precursor materials in forest plants, especially those of the tropics. If these plants are characteristic of the mature forest then extended felling programmes will rob them of their habitat at least until succession has built up the ecosystem again to something approaching the preceding condition.

Animals in the forest may be viewed in two different ways. They may well be a source of food for the local inhabitants and indeed the main source of animal protein in some places. Such food can be an important element in the diet of, for example, shifting cultivators whose main carbohydrate source is tubers and roots and in any case meat is enjoyed by almost everyone. Animals may also be conserved as wildlife for its own sake or as part of a tourist-oriented economy. (Forest animals are less popular in tourism than grassland creatures, of course, since they are less easily seen.) Not only may the mature forest ecosystem be needed but a large enough area of it is also required, to contain appropriate amounts of food, shelter and space for the species' social behaviour patterns. It may be very difficult to make these requirements compatible with extensive felling programmes. The general run of outdoor recreation activities (walking, scenic drives, picnics, wildlife observation, copulation) can be combined with extractive use since the recreationists can be channelled into particular areas, although they may protest at losing favourite views and at being excluded from reforestation zones (Plate 2.11).

Forests often have a creative role in watershed protection, principally in anchoring soils which would otherwise yield high densities of silt to the runoff. In mountains especially, they may delay the melting of snow and thus allow a steady release of water to the rivers. On the other hand, forests transpire a great deal of water and still more evaporates from intercepted rainfall on their crowns. Watershed management may therefore seek to trade off silt yield against water yield by keeping riverine areas forested at all costs while not objecting to logging further away from the streams. On steep slopes, however, most logging turns out to be environmentally damaging and sometimes irreversibly so. In the Phillipines the loss of soil from slopes is only part of the story. Every $400 \, t/km^2$ of silt in the offshore waters decreases the coral cover by 1% per annum; for each percentage point of coral loss, fish biomass decreases by 2.4 per cent. Animal energy and protein sources are thus diminished.

The inference we may draw from these paragraphs is that there are many multiple-use compatibilities but the match is not total. A key role is played by the type of logging used: various combinations of selective felling, clear-cutting, careful extraction and destructive extraction are all possible. The ecology of clearance is complex and varies with forest type but common elements will include for example the exposure of the soil to higher temperatures once the shade of the dominants has been removed. This may well speed up the activity of microflora and microfauna and accelerate the breakdown of organic material. A harder, 'baked' top layer may be one result, which will shed water much more rapidly than its predecessor. In régimes of intensive rainfall the silt yield may be high once the structure breaks down and everywhere any silt will carry with it quantities of nutrients. Such losses will only stop when the vegetation regenerates sufficiently to have roots once again in the upper layers of the soil. These can mobilise subsoil ions, scavenge nutrients out of precipitation and build up humus.

A different ecology results from intensive, energy-rich, forest replanting. Processes like fertiliser use, biocide spraying, ploughing, weeding, fire protection and grazing management (all of which may be partly or totally carried out from light aircraft or helicopters), are basically incompatible with other uses of the forest while they are taking place and in the case of the chemicals for some time afterwards. It is the less intensively managed forests which are the most capable of multiple use. Nevertheless, sustained yield and a degree of multiple use is possible in many forest areas, provided the scientific base is good and is accompanied by a management plan whose implementation can be properly brought about. Thus here, as in other matters, the developed nations tend to be the most effective where government and industry are concerned, though the involvement of grassroot movements in developing countries is becoming impressive.

Forest policies and futures

A summary of the world forest resource position centres on the following propositions:

- A decline in the area of tropical natural forests
- A shortage of wood in some low-income nations
- No general world shortage of wood and wood products

The areas of most rapid forest loss without reforestation are those in the poorer countries and appear to be outgrowths of rapid population growth and rural poverty. The woodland areas have declined under the impact of shifting cultivation, conversion to quasi-

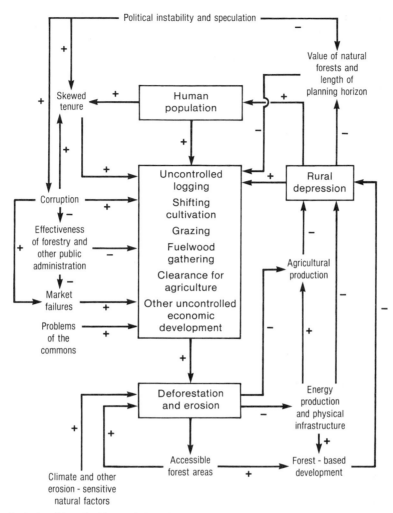

Fig. 2.10　A circle of interlocking relationships between population growth, deforestation and poverty ('rural depression') in the tropics. M. Palo, 'Deforestation perspectives for the tropics' in M. Kallio *et al* (eds) *The forest sector: an analytical perspective*. Chichester: Wiley for IIASA, 1987.

permanent agriculture, the gathering of fuelwood and the impact of large development projects. In addition, many tropical regions have experienced timber 'booms' with rapid exploitation of forests, which have been turned into non-renewable resources. These have been encouraged by investments that have opened up hitherto inacessible areas: mining, dams, roads and estate crops have all added to the availability of timber to logging companies. Government policies over taxes, credit and pricing that encourage private commercial exploitation on a short term basis have exacerbated the situation in many places. A conservative view also places blame on the common-property nature of many forested areas: in individual ownership, it is asserted, the need to secure the future would be perceived as much greater. Some of the many factors involved are summarised in Fig 2.10. Government policies may end up simply by banning logging. This may be difficult to enforce internally (as in Thailand after 1988) and also export the problem. Thailand's sawmill operators looked towards Laos, Kampuchea, Indonesia and even the Ivory Coast for supplies in the post-1988 period.

Indeed, it seems likely that many governments in areas of deforestation have adopted a rather skewed set of valuations for forests. The net benefits of exploited forests, for example, have been overestimated because many of the costs have been externalised and ignored; tropical forests have been used up in ignorance of their biology and the ecology of deforested terrain. The rapid utilisation of forest resources has been used to try and produce an income with which to solve fiscal, economic and social policies elsewhere in society. There has thus been a reluctance to invest sufficient resources in the forests for their efficient long-term management. In their race to appear modern, some governments have ignored the repository of traditional wisdom about the nature of the forest and its management, while not replacing it with imported science either. This has happened in many parts of S.E. Asia under the impetus of demand for Japan, which consumes 30% of traded tropical timber, mostly converted into plywood.

Tackling these problems in low-income countries is difficult, for even where there is no shortage of recipes, many of the desideradata for emplacing them are absent. Some involve action outside the forested areas, making difficulties even worse. The intensification of food production outside the forests, for example, would mitigate the encroachment of agricultural colonisation; such a process would be helped greatly by land reform but this is a political bomb in most nations. Some kind of Green Revolution would be helpful, in which genetically improved indigenous species were allied to investment programmes which favoured reforestation, and the rehabilitation of cut-over watersheds was allied to land policies which benefitted the landless. (The agricultural Green Revolution was lacking in this last respect.) The International Tropical Timber Agreement, ratified in 1985, contains the potential to encourage tropical nations to produce timber for industrial purposes on a long-term basis and might also be a help, though it is heavily criticised by environmentalist NGOs. In fact, as demand for wood grows in the world, the capacity of major suppliers like the USSR and the USA to expand their production further will be limited, and so there are large-scale opportunities for sustained income ahead and moreover ones which will create employment on a relatively large scale as well.

Discussion of forests nearly always centres around the actions of governments and it is true that in many nations the bulk of the forest land is part of the national patrimony. Nevertheless, local action can be significant: the role of NGOs and local associations is usually called community forestry and examples of its key role in managing forests and in reforestation projects have been found in Gujerat (India) and South Korea in recent years. We need also to recall the point above that land use policies outside the forestry sector are likely to be relevant as well.

On a world scale, variations in the futures of the forest resource are likely to come from a number of sources, such as rates of economic growth (and hence demand for paper products especially); variations in the exchange rates and of trade liberalisation which involved the disappearance of tariffs; the impact of acid rain in the developed nations; and the role of global warming, especially in northern timber-producing countries. Despite all these, and despite the fact that forestry lacks the grandeur of large dams, a resource use which has a future in stabilising carbon levels, keeping intact land and water resources, and providing energy and livelihoods in rural areas, cannot be overlooked. Nonetheless, poorer nations look to Europe and North America and see the prosperity engendered in lands where large areas of forest have gone and wonder whether they are being economically oppressed in keeping their woodlands.

Water

Satellite photographs of the planet Earth confirm that this is truly the water planet, that element giving it a delicate blue colouring. Water is in fact the commonest compound on Earth and is not only essential for life but has become interwoven with human livelihoods in various inextricable ways. Water is found on land, in the oceans and in the atmosphere, and comes in different forms: as a liquid, as a solid (ice and snow) and as a gas. It is also bound up with other molecules, as in living tissue. In free form it may be totally pure or may carry dissolved burdens of minerals, giving rise to the labels fresh, brackish and salt. It may carry many other substances in solution and in suspension, too. Because it occurs in very large quantities (perhaps $1.39 \times 10^9 \, \text{km}^{-3}$ is present on the planet), and is driven in a cycle by the energy of the sun, it appears to be a renewable resource and in absolute terms must be so since the hydrological cycle is closed on the scale of the globe. Nevertheless, as we shall see, human societies can intervene in the hydrological cycle in ways that diminish the renewability in practical terms. Lastly, we should note that in many of its roles in both nature and human activities, there is no substitute for water, as there might be for many other materials. When we use the term 'water resources', we nearly always mean *fresh* water, i.e with a low mineral, silt and biological content. This is needed not only for drinking and food processing but for e.g. industrial boilers as well, where a high mineral content rapidly leads to scaling. But such high qualities are not needed simply for carriage of vessels of or suspended wastes.

Water uses and needs

A list of the main uses of water by human societies is given in Table 2.19. It also gives examples of where substitutes are possible, and the percentage of supply that is consumed, i.e. used in the sense of being locked into some other substance (as in food processing for example where it might be put in cans) or not being returned to the source, as with irrigation water which is evaporated from the channels. In both cases the water goes back into the hydrological cycle but with alterations in time and space which we can call consumption. By contrast, floating some logs down a river does not consume the water at all. Water can also have its quality changed by use, as when run-off from an agricultural area contains pesticides which render the water unfit for humans to drink.

Emphases that should be drawn from Table 2.19 must begin with drinking water. Humans need about 2 litres/day of water which ought to be fresh water, i.e. lacking in appreciable quantities of minerals, free of suspended matter and as abiotic as possible, i.e. minus such life-forms as bacteria, viruses and the larval stages of intestinal worms. Other domestic uses include washing, cleaning the dwelling and food preparation, and as a carrier for sewage. Availability for these latter purposes is largely dependent upon a country's stage of economic development. Use per capita in e.g. Boston, Mass for showers, sewage and soups is 883 l/cap/day whereas in Nairobi the corresponding figure is 154 l/cap/day; in LDCs with standpipes the consumption is of the order of 20–40 l/cap/day. The second major emphasis is to note how few uses there are which have an acceptable substitute: in the case of ocean transport there is not even the land alternative that occurs with rivers, and the fact is that water is generally much cheaper than all its alternatives. In spite of high demand levels and competition for supply in rich nations like the USA, water is still inexpensive, as shown in Table 2.20, where its price for domestic and irrigation use is contrasted with cheap minerals. Lastly, the processing category of industrial use conceals the very large amounts used by industry: to produce a tonne of steel, for instance, consumes 8000–61,000 litres of water; a kilolitre of petrol (gasoline), 7000–34,000 litres of water, and

Table 2.19   Main uses of water

Use	Substitute	% consumed
Drinking	None	1–15
Other domestic	None	1–15
Public, urban	None	1–15
Livestock	None	1–15
Irrigation	None	10–80
Navigation	Land transport	0–10
Hydropower	Other energy	0
Mining	None	1–5
Industry		
Cooling	Air	0–3
Processing	Mechanical	0–10
Waste disposal	Air	0
	Mechanical	
Recreation	None	0
Flood loss reduction	Land use	0
	Management	

Adapted from A. K. Biswas (ed) *U.N. water conference, summary and main documents*. Oxford: Pergamon Press, 1978, p. 44.

Table 2.20   Comparative cost of water in USA, 1980s
US$,per tonne

Irrigation water	0.03
Municipal water at point of use	0.30
Sand and gravel (a)	3.00
Iron ore	30.00

(a) usually the cheapest mineral commodity
Adapted from D. H. Spiedel *et al* (eds) *Perspectives on water uses and abuses*. NY: OUP, 1988, p. 5.

a tonne of paperboard, 62,000–376,000 litres of water. Even minor consumer items float on a lot of water: a ladies' skirt takes 142 litres to make, and a pair of tights 0.8 litres. A tonne of worsted suiting has needed 266,000 litres for its processing: at 400 suits per tonne, every yuppie double-breaster has demanded 665 litres of water even before reaching the tailor's steam-iron. Thus in spite of the huge quantities of water present on the planet, it is not surprising that human activity can at times be a significant element in the flows of the hydrological cycle.

The hydrological cycle: natural and cultural

Driven by the energy of the sun, the world's water content fluxes ceaselessly between a number of reservoirs. At any one time, some 97% of that water is in the oceans and so it is marginal to the demands for fresh water. The portion of the flows that can be reached by humans centres on the phases in which water is precipitated in fresh form out of the atmosphere (where it has had a residence time of about 10 days) onto land (Fig 2.11) where it flows through the surface runoff channels like lakes and rivers, and the top few metres

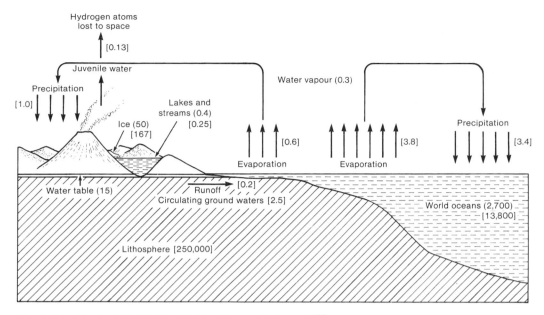

Fig. 2.11 The hydrological cycle. Numbers in [ ] are 10^{20} grammes, other numbers are estimates of the quantity of water present expressed as depth in metres per total area of surface of that type. P. A. Furley and W. A. Newey, *Geography and the biosphere*. London: Butterworth, 1983.

of the lithosphere, to the oceans where it is again lost as a freshwater resource. Residence time for a water molecule in this phase is about 17 days, so it has to be tapped on the run, so to speak. Of course, these aggregate data conceal a great deal of regional and continental variations, which are especially affected by the balance of precipitation and evaporation in a particular climate. Most living things are mainly water and altogether they sequester about 11,000 km³ of water at any one time, with a residence time of *ca* 17 days. In resource terms, therefore, the total volume of fresh water is less than 1/30th of the total global water content, and 22% of the freshwater is underground with only perhaps one-third of it at economic pumping depth. It is mostly the small (0.36%) but rapidly circulating proportion in lakes and rivers which humankind can evaluate as a resource (Table 2.21).

Intervention is however sought in other phases of the hydrological cycle. Cloud seeding with substances like silver iodide is attempted but with great statistical and legal problems in the fallout as well as any rain; desalination of ocean water is a staple source in places where energy has been very cheap such as Kuwait, in the USA (12% of world capacity in the 1980s) and on islands with a high population but little natural supply. From time to time, plans for towing icebergs from Greenland or Antarctica to Perth (Australia) or Saudi Arabia emerge but generally founder upon the economics of melt rates and tug-boat hire. If the technology is available and the price can be met, then quite deep ground water can be pumped up. In this way, deep aquifers can *de facto* be 'mined' since water is extracted far faster than recharge from precipitation takes place. The depletion of the Ogallala strata under the High Plains of the western USA is a well known case of this process. But it is in the surface runoff that most attempts to store water, divert it for various uses, to speed it on its way if there is a surplus (i.e. a flood) or pump it from shallow wells, all take place. While we think immediately of the large modern reservoir as a storage medium (Plate 2.13),

Table 2.21 Diversion of the hydrological cycle

	km^{-3}/yr	
Total annual global precipitation	52,100	of which
Falling on the oceans is	411,600	and
Falling on the land is	113,600	of which
Surface water run-off or heat-surface		
gd water comprises	41,000	but of this
The stable run-off (i.e. not floods) is	14,000	of which
Some falls in sparsely inhabited areas, ca	5,000	so
Available quantity for human use amounts	9,000	of which
Actual diversions for those purposes is	3,500	(38% of available) and
The total changed by human use is	5,800	

NB: Diversion as a % of total on-land precipitation is 3%; of global total 0.66%

Source: Adapted from data in P. R. Rogers, 'Fresh water' in R. Repetto (ed) 1985, 255–98.

most pre-industrial economies have had analogous if smaller-scale methods for storing water for agricultural and urban use; many are still in use. Once diverted, water has to be piped to its place of use and this may involve the use of energy if it has to be moved uphill beyond natural hydrostatic pressure levels. The windmill performed this service for drainage works in pre-industrial economies; nowadays a diesel pump is more likely. Energy is also embedded in the making of pipes and the digging of trenches and tunnels for them, and in for example the treatment of waste water. So water like so many resources can be tied to the energy fluxes not only of nature but also of mankind.

Supply and demand

These concepts are more difficult to deal with for water than most resources. Supply for example begs the question of how much, at what sort of quality, and at what cost in environmental alteration? Demand is equally fluid since it too presupposes particular levels of delivery, quality and price, all of which are mutable. Nevertheless we can extract some elements of the present and near-future situations and see what lessons they contain for water policy at various scales.

As a consequence of the cheapness of water and its bulk, there is practically no world market in water: local and regional availabilities are paramount in spite of heroic efforts in some countries to dam, pump and pipe on a large scale. Thus the access to freshwater enjoyed in different nations depends to a large extent on climatically-provided surpluses of precipitation over evaporation, on access to technology and on population densities and growth rates. Thus in the 1980s, the US figure for water withdrawals was 1986 $m^{-3}/cap/yr$ and that for Canada 1172, the USSR 812, China 460, Libya 408 and Malta 60, for instance. In Malta, all this water is used for domestic and public purposes, but the normal modal figure would be *ca* 10%. Industry has figures of the order of 7–15%, cooling of power stations 25% but the greatest water user of all, with a world withdrawal percentage of 73 is agriculture, notably where irrigation is involved. These aggregates conceal variations in supply of considerable significance, such as the fact that 3/5 of the population of the LDCs have no access to safe drinking water, and 3/4 no sanitary facility of any kind, water-using or not. Within the category of industrial use, a few industries account for most of the use: primary metal processing, chemical products, oil refining,

Plate 2.13 The Itapan dam under construction in the 1960s in Brazil. This was one of a number of large-scale schemes to produce electricity for industrial growth. The immediate environmental changes are obvious but the greater effects once the project is complete are not yet apparent.

pulp and paper and food processing account for 2/3 of the withdrawals.

Global estimates also conceal regional stresses in supply: Asia for example has 26% of global runoff (and much of that is of an unstable character) but 58% of population and so access is at less than 50% of the global average. Africa has equal amounts of runoff and population (11% of both) but the Zaire river carries 30% of the runoff through sparsely inhabited terrain and so 2/3 of African nations have below-average runoff for the continent and in recent years 20 African nations have experienced severe drought.

The elusive nature of water economics makes forecasting of future demand difficult but some ambitious targets have been set by the UN. For example, by 1990 it was proposed to bring safe water to an extra 768×10^6 rural people in developing countries and to extend sanitation to 322×10^6 people beyond recent levels, though water need not always be involved in that process. Aggregate demand seems sure to rise: users of standpipes in LIEs consume 20–70 l/cap/day whereas tap users consume 20–120 l/cap/day. So it is no surprise to see various estimates of demand for the future which exceed any realistic supply possibilities unless the price becomes very high. Given, however, population growth rates and the demands for industrialisation in the developing world, a rise in global withdrawals from 2530 km³/yr in 1970 to 3750 km³ in 1985 was to be expected, and the rate of growth will only slow down if there is greater efficiency in water use, especially by industry. Most nations of the developed world will in fact expect to see a decline in per capita usage in the years up to 2000 for that reason. Declines in LIE access are at heart likely to be driven by rapid population growth unequalled by investment in supply facilities and in some places exacerbated by absolute deficits caused by drought.

Too much and too little water can both cause considerable dislocations in human affairs and are often labelled 'natural hazards'. Floods and droughts are relative terms but both describe conditions outside the range of values of water presence (or absence) to which a society has adjusted. Thus the 1970s and early 1980s were an especially bad time for droughts in both the USSR and the Sahel zone of Africa. In the latter, the peak year of 1984 saw 30–35 million people affected, of whom 10 million were displaced. An average year in South East Asia sees the destruction by floods of 4×10^6 ha of crop land.

Adjustment to these hazards can be of four types. First, an attempt can be made to modify the event, as by tapping extra water supplies in the case of drought, or building flood-control dams on the upper stretches of rivers. Second, damage susceptibility can be modified like land use zoning to prevent investment on flood plains or adaptation of drought-tolerant varieties of crops. Thirdly, the loss burden can be lessened by the evacuation of people and expensive equipment or by remedial financial measures such as insurance. Lastly, the losses can be stoically borne.

Decisions about behaviour in these matters are primarily conditioned by economic considerations but these may contain all kinds of constraints: for example all the low-risk land may already be occupied by high-intensity uses so that high-risk areas are all that are available for economic growth. This is often the case in developing countries. For the rich, other measures may help: there may be a helpful administrative structure which enforces appropriate use or non-use of high-risk zones, or a river may have an integrated and on-line, real-time, computer based system of flood warnings, as have the Volga, the Tennessee and the Danube for example. Since economics, cognition of the hazard, and administrative structures are all so crucial, the term 'natural hazard' seems, in effect, to be inappropriate and it could well be replaced by 'environmental hazard', which leaves open the mix of the natural and the human which is nearly always involved.

Water and energy

We know already that water can be used to generate mechanical or electric power, and that it is needed in the production of commercial energy, often as cooling water in power plants. Here we shall briefly examine the obverse context, namely the need to use energy to provide water. This has been a feature of human life ever since Palaeolithic women carried up containers of the stuff from a stream to the camp, and Neolithic women hauled buckets of water from the well for daily use. These activities continue, but in modern societies the distribution mechanisms of water to households and to industry are rather more complex and involve commercial energy. Not usually counted in any energy budget involving water is the embedded energy in the plant: in the construction of dams, pipes and pumps, and in the digging of trenches and tunnels, for example. Future energy costs may well be affected by present activities: drawing down a ground-water aquifer, for example, to a level where bigger pumps and more power have to be used to tap it in years to come. Using water in energy production as in open-cast coal production and the utilisation of oil shales may affect the water quality to the point where it becomes a very scarce resource and thus puts limits on energy production.

Water management thus requires energy at most stages. Desalinating water is an obvious example for energy has to be supplied to do what the sun normally provides for nothing. Irrigation is another heavy consumer of energy, since it has to be used in the production of the technology, in lifting the water to its point of use and in applying pressure to spray the crops. Newer techniques like drip irrigation obviate the need for some of the energy use, but are often more expensive on capital equipment. In the USA in the 1970s, some 23% of all on-farm energy consumption was used in irrigation. For urban-industrial use, the energy costs of water supply can easily escalate: estimates prepared for Los Angeles, for example, show that ground-water (before it became saline) cost $500 \, kcal/m^{-3}$, the Colorado River aqueduct water was delivered at $1500 \, kcal/m^{-3}$, and the State Water Project at $2000 \, kcal/m^{-3}$. Desalination of brackish ground water needed $3000 \, kcal/m^{-3}$, of sea water $7000 \, kcal/m^{-3}$, and the towing of icebergs estimated at a mere $1000 \, kcal/m^{-3}$ ($1000 \, kcal = 42 \, MJ$).

Water and environmental management

Since water is crucial in the formation of so many phenomena, both living and non-living, it is not surprising that its diversion and use by humans should often produce environmental effects. If the scale of the diversions is small then the impact is unlikely to be great, but in the case of very large dams and their impoundments, and equally with management schemes for whole river basins, wholesale changes can be effected. The worldwide distribution of very large artificial water bodies is uneven, with Africa, the USSR and North America having the greatest representation; recent additions to the total have been in Brazil, Mozambique, Ivory Coast and Australia. The changes to the regional ecology are often profound, even if we omit the inundation of a large area of forest or steppe. The downcutting of the stream above the reservoir is reduced, so that aggradation may occur; below the dam the quantity of water, its chemistry, temperature and suspended load are all likely to be different with consequent effects on the geomorphology and biology of the river. For example, the river may not now have the power to move very much sediment, but its tributary streams may still contribute large quantities, producing something of a pile-up. Social effects may result from the drowning of villages and communications (and not only in developing nations), and the extension of irrigation may well bring with it a suite of diseases. Given the scale of many of the impacts, and the uncertainties over the real rates of net return on the huge investments needed, major world financial institutions have now decided not to bankroll any more of these projects.

Integrated river basin management is also environmentally manipulative since it nearly always involves large dams. In the USSR, the Volga River scheme generates $12 \times 10^6 \, KW$ of electricity, provides 50% of the inland fisheries of the nation, supports 75% of the domestic waterway traffic, waters 0.5 million ha of irrigated land, and transfers water to the Ural River, the Don and the Sea of Azov. Since 1972, waste control levels have also been integrated into the scheme. Many analogous projects require international co-operation and here the UN has often been a facilitating agency. In the basin of the Senegal River, for example, Guinea, Mali, Mauritania and Senegal have jointly agreed an action plan which coordinates a multipurpose dam in Mali with a saltwater exclusion barage at Diama. Many similar ideas have been mooted for other international basins, the Mekong being one of the most far-reaching.

Because water diversion can be costly in environmental terms as well as by purely economic criteria, conservation of water supplies by increasing the efficiency of their use has grown in popularity in recent decades. Since agriculture accounts for 70% of total world withdrawals of water, raising the efficiency of irrigation by 10% would in fact free enough water, if it were appropriately distributed, to supply the world's total residential use. New technologies such as the lining of channels in gravity-feed systems, drip feed from porous pipes, precision dropping from former spray arms, and better calculation of crops' water need, can reduce water demand by 20–60%. But on-site management is still critical: the range of improvement possible by increased attention to water conservation is wide.

For gravity feed systems the range is 40–80%, for centre pivot sprays 75–85% and for drip feeders 60–92%. In Israel, the world leader in this type of conservation, the volume of water used per irrigated hectare declined by 20% 1967–81 and made possible the expansion of the irrigated area by 39% with only a 13% increase in withdrawals.

In industry, re-use is often possible. The example of the cooling towers at power stations, which cut withdrawal needs to about 1/50th of those of once-through systems, is well-known. But many industries can design water re-use into their operations, for steel production can take 200,000 l/tonne or 5000 l/t; paper 350,000 l/ton or 60,000 l/t. Again in Israel, water use per unit value of industrial production has declined by 70% in the last 25 years. Household use is not immune from the capacity to change: this use is typically only 15% of a nation's water budget but it is an expensive sector to deliver and so lower use means less capital and less energy use as well. Toilets in the USA typically use 18–22 litres per flush, West Germans can make do with 7.5 litres per flush and Scandinavians 6 litres. Most household appliances can be designed to use 25–30% less water than conventional models. The re-use of waste water from urban areas is also essential in water deficit areas like Israel. The Dan scheme for 1 million people round Tel Aviv and Jaffa distributes 274,000 m^{-3}/day of partially treated water to ground-water storage and partial purification. This scheme aims to return 80% of the region's waste water to productive use. Further planning for the nation has the supply of 16% of all Israel's needs from treated waste water as a target for the year 2000 (Plate 2.14).

If wastes are not removed from water they may reduce its value as a resource and also have an effect upon other resource-environment systems, as discussed in more detail in

Plate 2.14 Most of the waste-water from central Israel flows not into the sea but through this plant at Rishon Letzion where it is treated so as to be available for immediate re-use in urban-industrial as well as agricultural contexts.

Ch 8. Here we may mention again that the mere act of developing water resources in some developing nations has increased the spread of a set of diseases, particularly those associated with irrigation which provides a good habitat for many disease vectors. Schistosomiasis affects 250 million people in 70 countries (about 7% of the human population) and filiariasis another 250 million, with 80% of the population in some areas having the condition of elephantiasis. Malaria and yellow fever are spread by mosquitoes, river blindness by flies and paragonimiasis by snails, as with the first two examples. Poor water management spreads the diseases of faecally contaminated water, such as cholera, typhoid, dysentery and amoebal infections, to which 1.5×10^9 individuals in the LDCs are exposed. Water, it seems, is best taken well diluted with an antiseptic agent such as alcohol.

Policies and futures

The general features of the water economy, which feed into any consideration of policies and the future of water resources, centre around a series of relatively simple ideas. Implementation may of course be rather less easy. The ideas are:

- The augmentation of supply by transferring water in space and time to meet anticipated withdrawal needs.

- Increased use of return flows: the highest quality needs are satisfied first, then others in decreasing order of quality need. Reduction of contamination assists this process.

- The reduction of rates of water use by making economies, by using new technologies and by the imposition of legally enforceable policies.

- Allocation of scarce resources in such a way as to minimise impact of shortages and the eradication of the notion that the supply of good quality water does not entail any cost.

Water allocation policies on a national or regional scale have been neither uniform nor static. In many legal systems, the riparian doctrine has been paramount, with the owner of contiguous land having the priority over withdrawals; prescriptive right is also common, depending upon a long period of actual and uninterrupted use. In drier areas, like Australia and the western USA, an appropriative doctrine came into effect which was based on 'first come, first served' provided the water was put to beneficial use. Modern resource management seeks to transcend these rather narrowly-focused allocation methods by looking to integrated river basin development and multipurpose use (Plate 2.15). If several uses are sought for a body of water, then various allocative problems have to be tackled. For example, collective benefits and costs have to be evaluated for a group regardless of individual preferences; such an evaluation will have to address the issue of intangibles, such as wildlife and aesthetic pleasures. The intricacies of water flows, too, are such that one usage is very likely to generate external costs for another use: heavy recreational use for example may spoil water for drinking unless treatment is then applied. There are also opportunity costs, as with the income foregone from farmland drowned for a reservoir. Lastly, there are always uncertainties and risks, such as unknown levels of future demand and supply, and fluctuations in the interest rate on borrowed capital. Nevertheless, water in DCs tends to be very cheap, compared with other essential materials (Table 2.20).

Techniques for helping with such complex decision-making processes include cost-benefit analysis which is based on the balance of the two elements in so far as they can be measured in cash terms. It is not without its difficulties, especially where intangibles are concerned and in some nations there have been suspicions that water development agencies have manipulated the inputs to get the right answer. Systems analysis has had its

Plate 2.15   Flooding is another challenge to water resources management. If not adapted to, severe flooding (as here in the São Francesco Valley of Brazil), lives and livelihoods are lost. In developing countries, any form of insurance against such calamities is unlikely.

enthusiasts but the equations needed to simulate the interface between physical and socio-economic processes are complex enough to deter many and they also forbid intelligent interest by lay people and so public opinion cannot easily be sought.

At the global scale, the UN formulated a set of priorities for the period 1977–90 which covered practically every water molecule on the planet. The list of headings for action is highly comprehensive and is followed by regional recommendations and an action plan with priorities. This latter focuses immediately on the need to extend outwards from the DCs the provision of safe drinking water and adequate sewage disposal, the shortage of trained people in the appropriate fields, the lack of suitable small-scale technology, and education in basic water-related hygiene. The actions called for involve massive resource transfers and it is not clear whether the interests of the developing countries may not be better served by rather smaller-scale approaches, as in the 'bottom-up' school of agricultural development.

Out of this and other reviews, a set of policy issues emerges. We can identify certain strands common to many parts of the world:

- The collection of data. This is often spatially thin and poor in quality, with error factors of hundreds of percent.

- The formulation and implementation of policy. The central government of a nation is the obvious body to do this, but some DCs prefer to leave the job to private enterprise and keep only minimum regulatory powers themselves.

- The setting up of physical planning units. The drainage basin is the obvious unit but this can be difficult because it may cross political boundaries, including international frontiers.

- The formulation of an adequate body of water law, to ensure economic and equitable water use. This will need therefore to extend beyond the water itself to embrace for instance land use, industrial effluent and toxic waste production, recreation, and wildlife interests.

Given rational approaches to most of these issues, water does not appear to be a global problem of the first order. There is still a great quantity of fresh water which is either untapped or which can be re-used if it is well managed. The difficulties are regional: in the DCs they are mostly of the treatment of wastewater and the unrealised potential for savings in withdrawals thereby conferred, and in the LIEs of supply for basic needs. Where problems exist, their solutions appear to be mostly within the economists' frame of reference for the difficulties of the LIEs may mostly result from an insufficient attention to external costs and those of the LDCs from far too little investment.

Estimates of global climatic change due to human activity are not yet sufficiently accurate to use in long-term water resource planning. But global warming would without doubt have a significant impact upon patterns of precipitation and hence of soil moisture and river flows, for example. The sensitivity of water resource availability to such changes will of course be highly variable in both time and space.

Renewable resources of the oceans

Unless we eat fish reasonably frequently, we tend to overlook the presence of the world's oceans in our recall of connectivities: any streaks of light are most likely to be concerned with our immersion in it than our consumption of its products. Yet the oceans cover $3.6 \times 10^8\,km^2$ of the Earth's $5.1 \times 10^8\,km^2$, which is 71%. The average water depth is 4.6 km, though most land masses are fringed with a continental shelf about 200 m deep. This gives a total water volume of $1.37 \times 10^9\,km^{-3}$, with a further $29 \times 10^6\,km^{-3}$ locked up as ice and snow, which could in theory melt and add to the oceans' volume. In the Pleistocene, there was an additional $44 \times 10^6\,km^{-3}$ of ice and snow, which is equivalent to 100 metres of sea-level. The water of the seas is characteristically salt, meaning that it has an average of 35% of minerals, a concentration that is lower near the outfall of big rivers (from which it receives 90% of the $25 \times 10^9\,t/yr$ of mineral matter which is the cause of the salinity) but higher in enclosed seas with high evaporation rates, like the Red Sea at over 40%. The chief constituent of the saltiness is sodium chloride which if extracted and spread over the dry land would form a layer 150 metres deep; magnesium would do the same to 60 metres.

As a habitat for life, the sea has temperatures from $-2°C$ to $+30°C$ but sunlight useful for photosynthesis rarely penetrates beyond 55 metres. Thus the continental shelves, which are closest to the sources of mineral nutrition, and upwelling zones which bring up mineral elements from deep water, are the zones of highest productivity: offshore the primary producers (mostly phytoplankton) may fix $50-170\,g/m^{-2}/yr$ of carbon; in upwelling zones this may be $1825-3650\,g/m^{-2}/yr$. These contrast with $50-100\,g/m^{-2}/yr$ in open water, which is equivalent to the terrestrial deserts. The absolute total of carbon fixed is estimated at $20-60 \times 10^9\,g/yr$ of carbon: about the same as the land surface. But given the large area of the seas, it is apparent that they are no obvious cornucopia for life: dissolved oxygen, for example, is about $9\,ml/l$ in the seas, compared with $200\,ml/l$ in air, which may limit some forms of animal life.

Ecology of the oceans

It is less easy to adapt the ecosystems of the sea to human ends than those of the lands: it is much more difficult for example to bring to bear high intensities of energy use via technology. Thus we need to understand the nature of these systems particularly well, yet it is obvious that they are the most difficult ones to investigate. Given that the seas' most important yield for humans is fish, then the food chains that support those animals are critical and they must begin with the fixation of solar energy by photosynthesis. The growth of phytoplankton is also limited by nutrients and so the zones where these are found are the areas of highest biological productivity. Thus currents, tides and winds which bring cold but mineral-rich water near to the surface are important energy inputs into the marine food chains. In fact the input of energy from these sources (and from rivers) probably overrides light and nutrient concentrations as determinants of NPP once into the open oceans. So NPP rates reflect continental shelves and upwellings where nutrients come up from depth and can be fixed by the phytoplankton. The food chains in the open oceans are very long, since the initial plants are very small; a 5-step chain (phytoplankton-zooplankton-small fish/larval stages-medium fish-large fish) is often postulated (Fig 2.12). In shallower waters, the phytoplankton is larger and chains of only 3 steps or even 1 are more common. The efficiency of the transfer of energy between stages is thus crucial in determining the populations of the larger fish which are so often the human resource; in some chains it appears that almost all the phytoplankton is eaten but in others a 20% efficiency is the best that can be achieved. So as Table 2.22 indicates, calculations of the total fish production (meaning biological production, not catch) can be made, which indicate a total fish (wet weight) of about 240×10^6 t/yr, of which only 1.6×10^6 t/yr is in the open oceans. The great importance of the continental shelves and the upwelling zones is thus confirmed and we shall need to remember this also in the context of pollution as well as of the oceans as a sink for carbon dioxide.

Living organisms as resources

The life of the sea and its salty margins have an age-old appeal to humans, for remains are found in most appropriate archaeological deposits. The main use, naturally enough, is for food since many marine plants and animals are edible. This is complemented, though, by examples of the use of for instance seaweeds as agricultural fertilisers, oyster shells for lime mortar and seals for their skins. Like many other resources, those of the oceans have passed through a number of pre-industrialised phases before the present day, though most of these have been variants on hunting and gathering, with the analogue of agriculture being very restricted until the present century. Through human history, fish have been the most actively sought and they have had the fortunate habit either of swimming in shoals or lying on the bottom. These dispersals meant that in pre-industrial

Table 2.22   Biological productivity in ocean zones

	% area	NPP/yr gC/m^2	Fish production 10^6/t/yr
Oceans	90	50	1.6
Continental shelves	9.9	100	120
Upwellings	0.1	300	120

Source: J. H. Ryther, Photosynthesis and fish production in the sea, *Science*, 166, 1969, 72–76

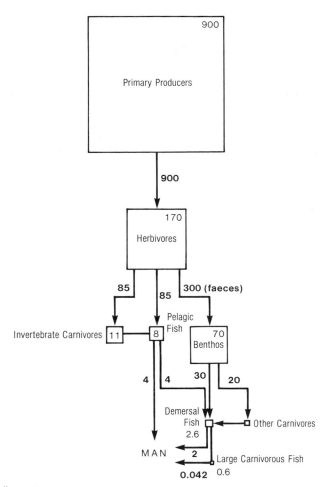

Fig. 2.12 Kcal/m²/yr of biological production in various compartments of the ecosystem of the North Sea. Note how long some of the food chains are. T. J. Pitcher and P. J. B. Hart, *Fisheries ecology*. London: Croom Helm, 1982.

times it was virtually impossible to cull a fish stock below its replacement level, and many populations lay beyond the reach of fishing vessels. Whaling, on the other hand, was carried on unto the ends of the earth but many stocks showed declines in catch or in the size of animal caught before the industrialisation that started in 1873 with steam vessels and explosive harpoons, a portent for the future.

The recent trends in marine fish catch are shown in Fig 2.13. This includes approximately 11% of the total from aquaculture but does not include the less conventional animals such as krill, octopus and squid, which total perhaps another 3.6×10^6 t/yr; neither are artisanal fisheries (used only to feed people at the landing sites or nearby), which might add 24×10^6 t/yr to the overall total. The trends show steady increases between 1958 and 1971, with a dip in 1972–73 which was caused by the collapse of the very productive Peruvian anchoveta fishery, but which then recovered to some extent up to 1986. The North Pacific pollack was also fished increasingly, adding to the increases up to the time

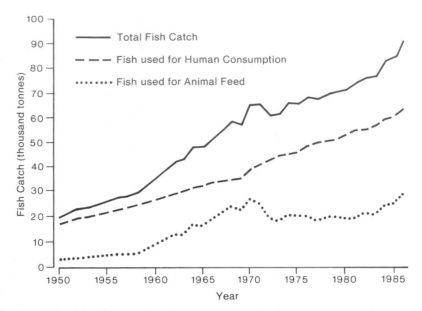

Fig. 2.13 Trends in global fish catch and its consumption by humans and as animal feed, 1950–85. *World resources 1988–89.* NY: Basic Books Inc, 1988.

of the latest data. The FAO keeps records of 280 fish stocks from its 19 designated world fishery areas and calls only 25 of the stocks either underexploited or moderately exploited. Thus the oceanic fish resource appears to be reaching something of a ceiling: FAO estimate that the maximum sustained yield of the global oceans may be about 100×10^6 t/yr.

As with most resources, there are regional variations in production and consumption. Asia dominates the world catch, with about 40–5% of the take, followed by Europe and the USSR. Canada heads the league table of exporters, followed by the USA, Denmark, Norway and Japan. Importers are led by Japan, with the USA, France and UK following after. For some nations, fish provides a very significant proportion of total protein: in the mid-1970s, for instance, the proportion in the Phillipines was 23%, in Vietnam 14% and in Hong Kong 18%.

The search for new or cheaper sources of marine food has led in one direction to little-utilised taxa. Squid are frequently touted as being capable of doubling the entire landing weight, though cultural demand is rather more sluggish than that; at one time the krill (a small shrimp) of the Antarctic was heralded as a great new source of animal protein and indeed Japanese and Soviet Union boats increased their cull of this species from 1000 t/yr in 1973–75 to 183,000 t/yr in 1983–85. It has to be processed within a very short time of catch, however, and so has a limited appeal outside those nations where animal protein can command high prices. The second direction is aquaculture, which represents a move towards herding of fish and other aquatic creatures (like oysters and mussels) if not actually agriculture, since the genetics of the species are rarely under control. In the west, most of the aquaculture is of delicacy fish like salmon and trout, which are expensive partly because they are high up food chains. In Asia, species lower down grazing chains are used, such as shrimps and milkfish, and in unpolluted coastal waters, the cultivation of detritus-feeding sedentary molluscs such as oysters and mussels is widely practised. Yields can be very high: the North Sea fisheries might yield 5 kg/ha/yr of fish and the Peru upwelling

60-5, but milkfish ponds average at 775 and oyster culture can reach 10,000 kg/ha/yr: compare pigs at 500 and cattle at perhaps 100 kg/ha/yr of meat.

If there is a slack in the world fishery resource system, then it comes with the quantity of fish that are caught to be converted to fish meal. At one time in the 1960s and 1970s this was seen as a cheap way of converting culturally unacceptable fish ('trash fish' as the trade unflatteringly calls them) to a tasteless additive that could fortify cereals and tubers in areas of nutritional stress. This has not come about, partly because the demand for fish meal as an animal fodder in the DCs has continued to hold up the price. As Fig 2.13 shows, however, the proportion of the world fish catch devoted to conversion to meal has risen from 20% in 1950 to 31–5% in 1986. While this goes mainly, like much grain, to feed the already well-nourished, then the rational use of the fisheries of the world is still some way off.

Energy and ocean resources

If in our minds we link energy and the oceans, it is likely to be in the direction of how to get energy out of them rather than energy expended. Yet in the course of extracting resources from the oceans energy must be expended just as it is on land (Plate 2.16). Even transporting people by sea has its energy costs once sailing ships have been superseded: a bike consumes 20 kcal/passenger-mile, an inter-city 'bus 140, a car 650, a jumbo jet 870 and an ocean liner 3125 kcal/passenger-mile[1]. For freight, however, railroads and waterways cost 170 kcal/ton-mile, a truck 700 and airfreight 10,600 kcal/ton-mile [2].

Just as we discussed the input of energy into land-based food systems, so we must note that for fish. Data generally only exist for the fishing part of the process, but we can imagine that once on land they are little different from other foods in the energy they need to bring to table. Statistics for the US fishing industry in the mid-1970s measure fossil fuel input per protein energy output (protein averages at 9% of wet weight of fish), in kcal per kcal. For boats over 5 GRT, the figure was 14, for those under 5 GRT 37 kcal/kcal. For comparison, feedlot beef took 20–44, broiler chicken 22, vegetables 2–4 and grain 2–4. So big boats consume less energy in getting protein than feedlot beef, but small boats are inefficient. If the marginal use of fossil fuels were being considered, then it would be a legitimate question to ask whether capital-intensive and energy-intensive fisheries were the best ways of getting human food from fossil fuels.

Other renewable resources

The most obvious of these is flotation in the sense that shipping does not consume water. It may however render the water unfit for other purposes. Wastes from vessels may massively interfere with the ecology of the seas and shores in the case of large oil spills. Less dramatic but more common is a gradual accumulation of oil residues in the seas (Fig 2.14) from on-shore tank leakages and other operations and from tanker cleansing at sea. This latter is mostly forbidden but still occurs. Other rubbish dumped in the sea, especially plastics, are also non-degradable and form much of the material along tide-marks today. Navigation, too, is not a matter of total freedom since, apart from natural hazards, there are other impediments to free passage. The development of off-shore oil and

1. These units are self-consistent, but for the metric enthusiasts, they approximate to 1–3 MJ/passenger-km for the bike and 210–3 MJ/passenger-km for the QE 2 or similar.
2. Equally true, but the equivalents are 11–6 MJ/tonne-km (rail and water), 47–8 for the truck and 725 MJ/tonne-km for the 747.

Plate 2.16   Though still hunting rather than agriculture, modern fisheries are linked to industrial energy flows, especially via oil: here needed to build the vessel, power it, provide refrigeration for the catch and unload it. With the addition of sonar and satellite data, the impact on fish populations is often severe.

gas fields for example may interfere with traditional fishing patterns as well as provide obstacles to all kinds of vessels. For military reasons, nations may declare exclusion zones. So although the high seas are not consumed by ships, their use is subject to a considerable number of constraints.

The other great resource of the seas is water. In its saline form it is little sought by human societies except for being near and for bathing. Desalination is possible using a variety of methods (Fig 2.15) but most are either very heavy users of fossil fuel energy (and hence mostly confined to producer nations) or else cannot produce the volume of fresh water that would make them serious suppliers of agricultural, industrial or municipal quantities.

At one stage, eyes were cast at ice-bergs and the towing costs of getting them to Los Angeles or Perth (Australia) investigated but in general the effort of re-using already existing supplies seems less. Equally, it may be easier soon to tailor crop plants to saline water than to desalinate the seas and so the cognition of the oceans as a vast tank of potential freshwater seems at present to be fading. Should nuclear fusion using deuterium (a hydrogen isotope, H-3) ever come about, then the seas will be the main source of that raw material and since it is present in very large quantities and renewed from the land, it can be regarded as a renewable resource; however anything but experimental demands seem at least 30–40 years off.

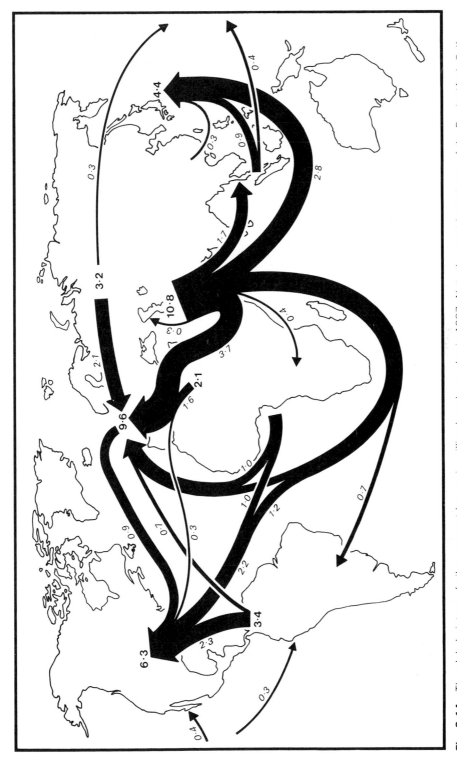

Fig. 2.14 The global picture of oil movement by sea in million barrels per day in 1987. Note the great importance of the Persian/Arab Gulf. Clark, R. B., *Marine pollution*. Oxford: Clarendon Press, 2nd edn 1989.

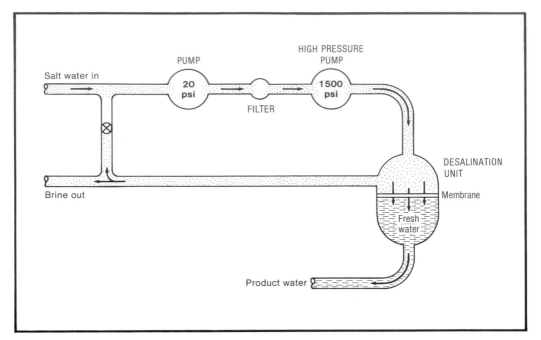

Fig. 2.15   The flow of water through the reverse osmosis desalination process: 'product water' is the fresh water output. Deming, H. G., *Water, the foundation of opportunity*. New York: OUP, 1975.

Fisheries management

At a pre-industrial level, the impact of fishing upon the populations of target organisms was sometimes felt. Either a mechanism for restricting the catch was devised locally or the fishery collapsed and the stocks then built up again naturally. In Oceania, several traditional practices evolved to ensure a continued yield from coastal sources: some of them deliberate and some inadvertent: specific species regulations, taboos, closed seasons and food avoidances were all found.

At a biological level, fish populations are characterised by the large number of them which swim in shoals and by their reproductive habits of spawning large numbers of eggs which lead to high quantities of juvenile stages well below any economic threshold but in which there is a very high mortality. Many populations seem to undergo cyclic fluctuations in abundance under natural conditions. Sea mammals, on the other hand, bear few young and at infrequent intervals but parental care ensures a higher survival rate.

Onto these types of ecology has been grafted the modern fishing industry: some of it artisanal, catching for 'local' consumption but much of it fully industrialised including freezing and processing at sea, as well as engaging in on-shore processing like any other part of the food system. Better technology constantly improves the ability actually to capture fish: electronic navigation and sonar location, nylon nets and vessels capable of long voyages to hazardous seas are the most obvious sources. Thus ever since the 19th century some fish stocks have been in peril of overfishing, and maps exist showing how many started to collapse once steam vessels were introduced (in which development of the steam-powered net winch was every bit as important as the main engine), a trend that has continued ever since, with the occasional interval for war. So many stocks have at one time failed to recruit adequate numbers of young animals to replace those taken from the breeding population and the industry has had to take smaller fish (which may exacerbate the problem), move to different species (which may be of lower value), or contract. As is well known, this sequence has applied *a fortiori* to whales, where there has been a shift from large whales to small and eventually in 1982 (extended for another year in 1990) to a moratorium on commercial whaling, though probably too late to stop a number of species from biological extinction. Those not made extinct may be at very low levels. There are now perhaps 200–1100 blue whales left, compared with 250,000 a century ago. Fin whales were estimated at 100,000 in 1975 but now appear to number only 400. The Antarctic minke stock, subject of 'scientific' whaling by a few nations (including Japan) is 440,000–690,000 strong. Some fisheries also kill large numbers of non-target species

Plate 2.17 A Japanese driftnet in the Tasman Sea in 1990 has ensnared a Sunfish together with a smaller species. Neither is the target group.

which are caught in the nets, die and are then dumped or sometimes illegally sold: some of the tuna catchers of the Pacific for example, are responsible for the death of very large numbers of porpoises (Plate 2.17).

Fisheries management, then, has attracted a good deal of attention from both biologists who wish to conserve long-term stocks and economists who wish to marry that aim with the objective of a profitable industry in the short-term. Other objectives may be the continued livelihood of those engaged in the industry itself, which like others of a hazardous nature (deep coal mining, for example) invokes a loyalty to a way of life which is incomprehensible to outsiders.

Fishery management can of course be left to a free market on the grounds that an over-exploited species will become uneconomic to take long before it becomes biologically extinct, which is probably true of fish though less so of whales. There are likely to be socially undesirable consequences of such a policy, and it is also deemed to be an inefficient use of capital; further, a stock may not simply recover its former levels because its place in the food chain can be taken by another and uncommercial species.

So governmental regulation (and indeed international regulation) is widely practised in fishery management. There are many ways in which governments can try to regulate fisheries, most of them falling into the categories of gear limitation (e.g. net mesh size), mortality avoidance (e.g. closed seasons and zones, licences for boats, and catch quotas), taxation, and by the encouragement of alternatives such as fish farming. It is, of course, one thing to enact regulations and another to enforce them, especially in developing countries. Fisheries management, then, has suffered from a number of inter-related problems. Natural populations of fish are difficult to understand, let alone those heavily manipulated in the past, and schemes for regulation are almost invariably at odds with short-term pressures. There is often no clear agreement on the optimal outcome of any strategies and the economic and political costs of implementation are usually high.

For many years, attempts were made to define for each stock species a Maximum Sustained Yield (MSY, Fig 2.16) but this has more recently been seen to be a deficient concept. The values of MSY were often set too high because the populations were not in equilibrium, and in any case regulation of fishing effort is probably a better control mechanism than total catch. MSY, for example, gives no guide as to how to parititon effort or catch between units in the industry, and in the long-term it is an average value which is useless in the face of high year-to-year variations in stocks. The replacement concept is called Optimum Sustained Yield (OSY), which tries, ambitiously, to meld together biology and economics with the socio-political factors that are so important. The end result is held to be the maximum benefit to society as a whole; however such a broad concept leaves a great deal of room for negotiation of variables. Nevertheless, fishery science is beginning to implant the idea that different stocks may need very different management régimes; in some, recruitment seems to be driven by environmental factors (e.g. the persistence of an upwelling driven by winds that may not be constant every year) and in others, natural predation is the chief factor controlling stock levels. This second is clearly more stable and easier to manage but the first is capable of very high yields under some conditions. But industrial fisheries are without doubt in a position in which many of them barely make a profit and are kept in existence with government subsidies.

If OSY is a rather loosely defined set of aims, then one of the biggest changes to affect fisheries management in recent years is rather more concrete. This is the outcome of the UN Law of the Sea Conference of the 1970s that enacted a Convention that came into force in 1982 which created 200-mile Exclusive Economic Zones round coasts. These in fact transfered 99% of the world's fish catch into waters which are capable of regulation, rather than being part of the global commons. The biggest gainers were developed countries but

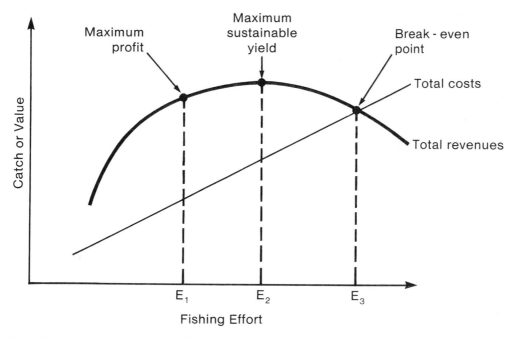

Fig. 2.16 A standard curve for MSY in fisheries. Note that maximum profit is in fact below MSY. (Fishing effort = number of boats times number of days at sea)

some LIEs in West Africa and the Pacific also gained. Given the ability to regulate, nations without large fishing capacities can now sell their surplus to other fleets.

The ultimate in management so far is fish farming which may focus on salt-water fish like salmon but also on brackish-water fish like the milkfish in Asia or on other classes of organisms such as oysters and mussels. This is the closest that the industry has got to post-Neolithic ways, though even here genetic control of the animals is rare. They are, however, confined, fed, culled at optimum weight, and treated for the many diseases that affect confined populations. In Asia, such fish are a very important part of the protein supplies of large numbers of people, whereas in the West the production of luxury items such as shellfish, oysters and salmon is more common.

Futures

The main problem of the seas as a global commons for fish has been addressed by the UN and with the creation of EEZs, the way is open for a new era of rational fishery management. An open way does not of course mean that everybody will follow it. A major interaction is the extent to which the global commons' use of the seas as a dump for wastes of all kinds affects, and has the potential to affect, marine populations. If poisoning of the oceans, which is clearly quite easy to achieve, becomes more widespread then it could have very serious effects upon a vital food resource; the vulnerability of marine organisms is obviously high since the food chains start with phytoplankton and they are very easily affected by toxic substances even at very low concentrations. This consideration has to be raised again when we consider the other great role of the oceans as a global resource for all mankind when they regulate climate. Part of this involvement in the great global distributions of water, temperature and carbon seems to be achieved through the presence of life. It seems that phytoplankton emit dimethyl sulphide (DMS), which in aerosol form becomes the condensation nuclei for clouds over the oceans. The higher the productivity of the plankton, the more clouds are formed and thus a regulation of temperature is achieved; secondarily the rainfall over the oceans and on down-wind landmasses is also affected. Thus the health of the oceans is just one more link in the life-support mechanisms of planet Earth which are crucial to us all.

Renewability and conditionality in renewable resources

In spite of the condensed and necessarily patchy accounts of resources given here, it must be clear that the concept of renewability of resources is a conditional one. That is, the resources are renewable provided either nature's flows are little manipulated or that a rational management is pursued which ensures the sustained supply of the desired materials. Many cultural practices can interfere with either to bring about, for example, a short-term high yield followed by extinction of the resource, or a prolonged attrition beyond the sustained yield level of the resource which makes its availability uncertain and hence expensive, or a toxification of environment which renders biological production or water quality to a nugatory level. At one extreme, the scenarios for 'nuclear winter' bring about this condition of very low biological productivity and it is this phenomenon, after all, upon which we all depend: it is the ultimate resource.

Sustainability is a word much used of renewable resources. But natural resource depletion is rarely entered in national income accounting. It is possible for a country to exhaust its minerals, erode its soils, burn and log its forests, grossly contaminate its air and water resources, and fish out its seas, and the way in which national income is measured would not reflect those changes. Among other problems, this means that there is a conceptual dichotomy set up between 'economy' and 'environment' as if they were unrelated. Clearly, change is overdue.

One characteristic of renewable resource systems is that many of them are still dominated by solar energy. The oceans, the hydrological cycle and forests, for instance, are still largely powered by solar input, and it is very much the food system which has swung to the other end of the spectrum, with DCs eating items which are mostly derived from diesel oil and in some cases rather taste like it. A very important question for the future is, which of these two categories ought we to emulate in the face of at least one more doubling of world population; ought we to try to convert the whole world to 'organic' forms of production or are better conditions for the poor to be found in the extension to them of the levels of energy subsidy found in the developed world of today, including the food system?

6

Non-renewable resources

Apart from objects ejected into space by modern technology, Earth is a closed system for materials. When, therefore, we talk of non-renewable resources we mean first of all materials which are so transformed by their use that they are not employable again by human societies. But in one form or another they are still present on the planet. There are in fact three main categories: those which are 'consumed' by use such as coal, oil and natural gas whose complex molecular structure is broken down into much simpler components; theoretically recoverable materials such as minerals, which are technologically capable of being recovered after use; and recyclable substances such as metals and glass which can be re-used without an enormous amount of re-processing.

In practical terms, materials such as metals, ceramics and plastics may become so dispersed by use that they cannot readily be sieved out again from the repositories in which they are placed: plastic bags and wrappings are an example. They may be physically impossible to recover since they are in a very dilute form and dispersed in water or air: the lead in aerosol form used as an additive in petrol is an example. At the other extreme they may be so concentrated that they are too toxic to handle at any rate for a long period: some of the wastes from nuclear power generation fall into this category. Lastly, the materials may be sequestered for such a long time that for all practical purposes they are lost to re-use: the steel frame of a large building locks up great quantities of materials. Certain patterns of use, too, can turn renewable resources into non-renewable ones: the human-induced extinction of a biological species is a form of irreversible consumption.

General characteristics of non-renewable resources, then, are that they are usually products of the lithosphere, that they usually need complex processing before use (with linkages to energy consumption and the production of wastes), that they enter world trade and so are moved around the globe and have been much more important quantitatively since the 19th century, and become depleted since so much of their use has been a 'once-through' process. This latter raises the question of the optimal depletion rate: should it emphasise the perceived needs of future generations and thus conserve the material as much as possible, or will we do the best for our descendants by using as much as we wish in order to turn it into knowledge of how to do without it? The complex field of resource economics is much concerned with this last question.

In such analyses, a fundamental question about any non-renewable resource is always, 'how much of it is there?' This is not a simple question, if only for the reason that exploration of the earth's mineral resources, for example, is not complete. Indeed, the amounts available in practice at a given time depend upon five factors as enumerated by Rees:

- The availability of technological knowledge and equipment and their location in the right places and amounts.

- Levels of demand, which encompass many constantly changing variables like population growth, affluence, tastes, government policies and the availability of alternatives.

- Costs of production and processing. These reflect the nature of the material and its location, the state of the art of production as reflected in its costs, including those of energy but also capital, the rate of interest on loans, taxation, and the risks of being nationalised or terrorised.

- End-price: this will reflect not only the factors above, but also pricing policies of the producers and government subsidies or taxes.

- The attraction and availability of substitutes, including the use of recycled products as against virgin materials.

Hence the resource is scarcely a fixed physical quantity (though this must exist) but a rather fluid economic and social construction. A common variable, for example, is price: as the price of a material increases it becomes more worthwhile for prospecting to take place, or for better methods of recovery to be developed. In this way, the recovery of crude oil from rock strata rose from *ca* 25% in the 1940s to *ca* 60% in recent years. These considerations have led to a conventional classification of earth material (especially minerals) of the non-renewable resource category. This uses the term 'resource' for the totality of the materials, but 'reserve' is employed for near-market materials. In turn, these are classified into measured, indicated and inferred groups, which are self-explanatory terms. The remaining 'resource' is either hypothetical or speculative, depending on an assessment of probabilities derived from existing knowledge. To be emphasised is the fact that the boundaries are always altering as the factors discussed above change. The actual quantities of some minerals derived from such an analysis are discussed later.

The main categories

Minerals are the core of non-renewable resources, which are usually divided into fuel minerals (oil, natural gas and coal), and non-fuel minerals. The former category have been dealt with and only the latter are discussed in this chapter. But we may also consider, albeit briefly, the land itself as a resource of this kind since it is very largely, though not entirely, a fixed quantity.

Land as a non-renewable resource

Each year, there are land gains and losses, with the portions of the surface of the earth becoming a resource in the sense of a usable surface, or losing that status. Coastal erosion and deposition are the most obvious categories, but landslides and soil erosion may also be found. Some of these changes are the result of natural processes, as when cliffs of soft material are exposed to high energy seas; others result from human activity as when for example coastal structures provide traps for silt and sand and thus build up ground above tide-levels. Occasionally, more spectacular losses occur, as when a volcano spews lava over former forests or cropland; the equivalent gains are made when a nation like the Netherlands dykes off large areas of coastal mudflat and saltmarsh for conversion to pasture and crops.

If land is in short supply, then the response of many societies with a choice is analogous to agricultural expansion. Option one is to intensify the use, leading often to multi-storey buildings in sought-after areas of cities; option two is to extend outwards by 'reclamation', which may mean many things but usually signifies the bringing into the economic framework of land whose benefits were formerly negligible like industrial wasteland, or not quantifiable, like coastal marshes. (This does not imply that these latter lands are not very good parts of the national patrimony, merely that in narrowly economic terms they are of very low monetary value.)

One exacerbation of land shortages is by degradation of the land surface to the point where it has little or no value. Dumps of toxic waste, for example, can have no other function since they are dangerous and sterile; land prone to subsidence (due to fluid withdrawal or mining) may have a few uses but if it is unstable then it may just be left; and even in planned land-use systems, there is planning blight in which land awaiting a change in function is idle and often grows nothing but weedy vegetation and discarded hypo needles.

Thus, although not strictly speaking, a classic non-renewable resource, land has enough in common with the others of that kind to merit mention: Mark Twain said that he figured he 'would invest in land, 'cause they ain't makin' it no more'.

Mineral resources

These are archetypal non-renewable resources in the sense that the deposits from which they are taken are formed on geological time scales of a totally different order from that of the human scales of their use. With the implantation of technology even in LDCs and remote areas, possibly 90% of the human population depends upon minerals not simply for industrial life-styles but just for survival. A DC inhabitant, however, may consume large quantities of minerals over his or her *ca* 70 year life-time: 460 t of sand and gravel for example, 99 t of limestone, 39 t of steel, 1.4 t of aluminium and 1.0 t of copper, are all typical figures, in this case for West Germany in the 1970s. About 100 non-fuel minerals are traded and contribute about 1% of world GNP; of these there are 20 metals of considerable importance and 18 non-metals of equivalent significance including aggregates, asbestos, clay, diamonds, fluorspar, graphite, phosphate, salt, cement, silica and gemstones. Importance is not always equivalent to quantity produced since some are needed only in small, though vital, amounts like steel hardeners such as tungsten and wolfram; presumably to have gemstones available at pig-iron ingot size would rather detract from their value. Further, although the use may seem humble it may be very important: consider the use of metals in all phases of the food system, for example, from machinery through to processing and packing. The end-point of this scale of values is the designation of some minerals as strategic minerals and the stockpiling of them against politically-induced shortage on world markets.

The resource flow

These resources are examples *par excellence* of the necessity for capital inputs to turn raw materials into useful substances (Fig 2.17). Only in the case of a few minerals like sand and

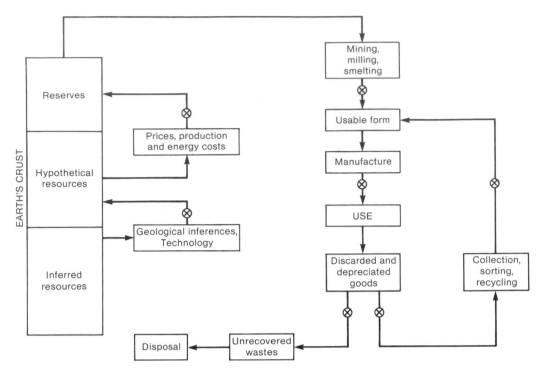

Fig. 2.17 The flow of minerals through the economy, with some of the major influences. The tap symbol represents a 'valve' where the rate of flow can be controlled.

gravel is the natural material usable more or less in the form from which it is extracted from the ground, with a bit of washing and sorting. Other minerals have to be subjected to a sequence of processes which begin with mining (Plate 2.18). In the USA over 85% of minerals are produced by open pit mining and this is probably a typical figure, with the rest coming from underground mining, solution mining (e.g. of salt), offshore dredging, and extraction from seawater by evaporation (salt again, but also aluminium). Then follows the process of benefication in which the mineral is winnowed from its ore by mechanical means which usually means crushing and sorting, the latter by a variety of methods. This is followed by refining, which aims to remove the impurities and covert the mineral to a form in which it can be used by commerce: smelting, electrolysis and hydrometallurgy are all commonly used.

The richness of the ores is very variable: haematite can be almost 100% iron but this degree of concentration is very rare: copper ores currently in use are more typical and they range from 0.1% to 2% copper, and naturally the richest ores have been used first (along with the most accessible) in the case of most minerals. The richness of the ore determines to a large extent the quantity of energy needed to produce the mineral for the market (Fig 2.18). Normally, energy input rises steeply and disproportionately as the quality of ore falls. Thus the price of energy is an ingredient in determining what grades of ore are exploitable at any one time, so that iron, aluminium and manganese are 'cut off' below about 30% for example, copper and nickel at 0.5% and gold at 0.001%. Clearly, energy is not the only factor but can rarely be ignored (Plate 2.19). Given moves through time to lower grade ores, it need not surprise us to learn that plants have got bigger: the talk of

Plate 2.18 An open cast copper mine in Queensland, Australia. This illustrates the impact of mineral ore extraction and reminds us also of the input of energy needed to take out (and then process) such large quantities of rock.

the industry in 1870 was the plant at Clausthal in the Harz Mountains which processed 500 t/day; now it would take over 100,000 t/day to keep the mining men from their beer. But the size of the plant may also be determined by the deposit: this is crucially limiting in both quantity of throughput and location: you cannot move the mine to another place. Other parts of the process can be moved provided that transport costs are low and so great bulk vessels and trains can transport minerals huge distances, like iron ore from Australia to Japan; all this in spite of the fact that mineral ores are often not high-value products.

Major global producers of minerals

Given that there are many usable minerals and many nations, a summary picture of production might be thought to be an especially difficult task. Yet if we look first at *reserves*, we find that 20 minerals account for 90% of the value of all reserves, and that they are concentrated in five countries: USA, USSR, Canada, Australia and South Africa. For 10 out of the 20 most important minerals, those lands contain > 75% of the reserves. In fact > 5% of the world reserves of all 20 are in the USSR, which has a significant supply of all minerals. After this remarkable concentration, a small number of African, South American and Asian countries account for most of the rest.

When we consider *resources*, the position does not change much, but there are two differences: first the great ignorance about China's resources, and second the magnitude of the sea-bed deposits ('nodules') which are rich in copper, cobalt, nickel and manganese and lie outside the 200-mile EEZs. One survey, for example, showed that an area of sea-bed in the Pacific some 300×2500 km might contain 2×10^{12} t of these nodules.

So *production* rates contain few surprises, and the concentration of the industry is

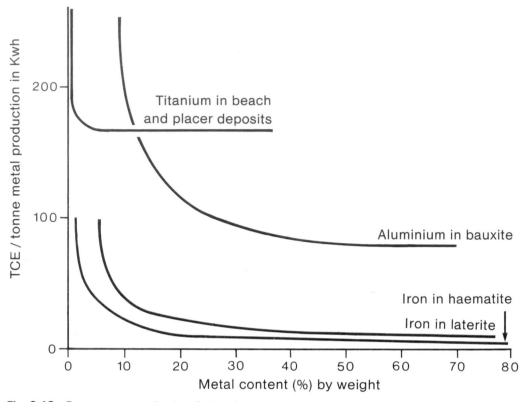

Fig. 2.18 Energy consumption in refining of metals. As the metal content of the ore diminishes so the energy (as Tonnes of Coal Equivalent) consumption to produce a tonne of metal increases rapidly. (Compiled from various sources).

maintained (Fig 2.19). In fact, 70% of the volume of the world comes from about 170 mines, mostly found in the ranks of the nations which dominate the reserve scene. Hence five nations (as above) produce > 50% of 2/3 of the major minerals, with Canada leading the world on a value per capita basis. Australia is the newcomer to the list but now gives us 20% of world output volume, and South Africa dominates the production of manganese, chromium and gold; of the others, China is first for the production of tungsten and antimony. The concentration is accentuated by the trend towards very large MNCs so that a few companies have near-monopolies of, for instance, molybdenum, platinum and diamonds. Their involvement with LDCs has led to the perception that LDCs are especially important as suppliers of minerals to the west and Japan but this is scarcely so, though Chile, Zambia and Zaire between them control a large part of the world's copper production. The LDC proportion of the global output of copper, bauxite (for aluminium) and lead has been decreasing in recent years and increasing in the cases of iron, nickel, cobalt and tungsten.

Consumption and trade

In the past 80-odd years, consumption of minerals has risen by a factor of 12 and we can say with confidence that since about 1950 the world has consumed more minerals than in

Plate 2.19 At this Welsh plant, steel is produced from raw materials by the most modern methods, using much lower unit quantities of energy and water than a decade ago. Such processes are subject to continuous improvement. Their landscapes are however stark and all-pervasive.

the whole previous history of humanity. The value of consumption is still rising, especially for metals (Table 2.23). Most of this has been located in North America and Western Europe (with Japan catching up fast) to the extent that the US economy demands 20 t/cap/yr of new minerals. The LIEs only take up about 10% of the total but their growth rates are higher at present than those of the HIEs. The non-energy uses of petroleum are encapsulated in the data for the rise in world consumption of plastics from 1.15×10^6 t in 1939 to 80×10^6 t in 1989.

Table 2.23 Regional metal consumption per capita actual and projected, 1975 $US

Region	1975	2000
USA	200	290
Western Europe	120	180
USSR & Eastern Europe	140	220
Latin America	30	40
Africa	4	7
Asia & Oceania	8	12
Other industrial countries (i.e. Canada, RSA, Japan)	180	320

Compiled from various sources

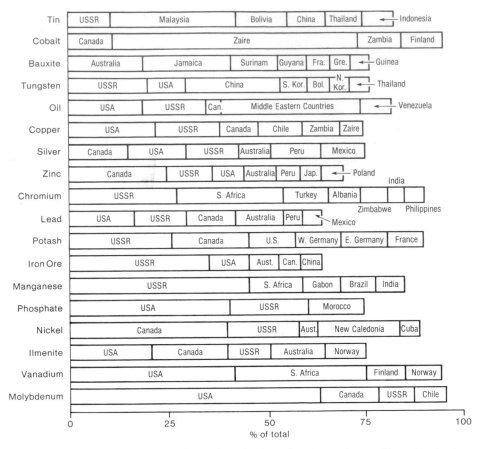

Fig. 2.19 Percentage share of world mineral production of major producers. Note the dominance of the USSR, USA and Canada. M. H. Govett, *World mineral supplies.* Amsterdam: Elsevier, 1976.

World trade in minerals in the past 20 years has exceeded GNP growth by a factor of more than × 2 and so Western Europe, Japan and the USA have come to be dependent upon imports; if fuel minerals are included then 40% of Japan's total imports are of minerals. Such trade patterns can mean that some LDCs with large deposits can come to rely on their export for foreign currency: copper comprises 95% of Zambia's exports and in Zaire, 67% comes from iron. With some concentrations, OPEC-type cartels are always possible and indeed likely given that LDCs usually see the terms of trade as being unfavourably stacked against them. Such factors encourage recycling in the DCs and so we note that 50% of British Steel's output is from scrap (Plate 2.20) and that in a nation like the USA 35% of lead is re-used and between 20–30% of copper, nickel, antimony, mercury, silver and platinum. Aluminium in the USA is a good example of the interaction of energy and recycling. Primary aluminium takes 45,000 KJ/kg to produce, whereas recycled material takes only 2000 KJ/kg, a saving of 95%.

The trends suggest a tension between more trade and less self-sufficiency in the consuming nations. The LDCs, Japan and Centrally Planned Economies are set to increase their consumption and in general this will lead to more large-scale production at the lowest possible cost except where environmental considerations are seriously taken into the reckoning. Exploration of less conventional sources will increase and there will be conflict over the exploitation of the resources of the ocean beds outside the EEZs.

Environmental considerations

The world over, the effects of mining and processing of minerals upon the environment cannot be ignored (Fig 2.20). In the case of the land, $2-3 \times 10^{12}$ t/yr of rock and soil are estimated to be moved in this cause and projections of current rates suggest that by 2000, some 24×10^6 ha will be affected, which is about 0.2 per cent of the land surface of the globe. Within these totals, experience in the USA points to some 60% of this disturbance being due to extraction, with most of the rest being used for the disposal of wastes and a mere 3% is land subsidence due to underground operations. For an individual mineral such as copper, the production in the US of 5.5×10^6 t of copper ore concentrates meant the mining of 245×10^6 t of copper ore, and the leaving of 240×10^6 t of tailings. The output of 1.6×10^6 t of blister copper resulted in 2.7×10^6 t of solid waste in the form of slag. The production of heaps and holes and poisoned land was very much a feature of the 100 years following the industrial revolution, though not uncommon before that on a smaller scale. Our attitudes now require attention to be given to the reclamation of mineral land, and one estimate has given the worldwide figure of 40–60% for such re-use. In the DCs large holes are always in demand near cities for the burial of urban and industrial wastes, and access to energy in those nations makes possible the levelling of tips and their conversion to all kinds of other uses: housing, forestry, agriculture and recreation are all possible depending upon the kind of treatments that are available and how effective they are in a given location (Plate 2.21).

Smelting of metals is a particular feature of mineral processing and one of the products is usually sulphur, a noted contributor to acid rain. A large copper smelter, for example, can emit 7400 t/day of sulphur if there is no treatment of the waste gases. Downwind from such plants there is usually a plume of affected vegetation, with very few living things in

Plate 2.20 These car bodies await another journey to be recycled and used again. Recycling of metals saves energy and raw materials but construction of cars to last much longer would lower consumption even more.

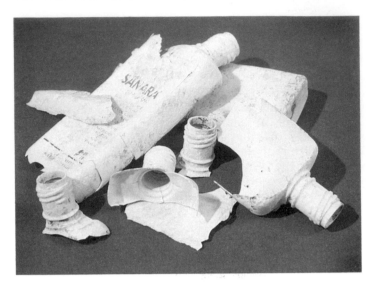

Plate 2.21 Plastics have been a major volume component of municipal wastes as well as of urban, riverine and marine litter. One reason is their longevity and here are examples of biodegradable plastics under trial by their manufacturers, ICI.

the areas of highest fallout; it is very similar to the vegetation pattern around sulphurous hot springs. Humans are also badly affected in such places: neither their bodies nor their possessions react well to high levels of sulphur, for example.

Rivers may also be affected by minerals extraction, when the plant manager is allowed to discharge waste materials into streams or when runoff from waste heaps enters the surface flow. Salt from the potash mines in Alsace is one of the noted pollutants of the Rhine, for example, and rivers in the Pennine districts of England still bear the effects of lead-rich discharges in the 19th century. Some minerals may build up in sediments in low-energy environments such as estuaries and thence via bioaccumulation become lethal to a wide variety of organisms. If rivers flood, then toxic metals for example may be spread over farmland; in the Phillipines some 130,000 ha of irrigated land are said to be affected by mine tailings that have entered the water distribution system.

In total, therefore, there are a number of different pathways in which metals, for instance, can be transported to even one pathway, such as fresh water (Fig 2.21). Add to these the formation of land surfaces (tips, fill, tailings) with high metal concentrations, and air contamination at many sizes of particle, and we have scope for considerable environmental impact. Once in fresh water, the form of the metal is very important in determining its effect on living organisms. This form may range from free metal ions through to precipitates and obviously brings about a variety of results, from the negligible to the blanketing of an entire stream-bed with red iron oxides.

An unknown area is the effect of deep-sea mining on ecosystems (Fig 2.22). If oceanic mining of mineral nodules for e.g. manganese and nickel is pursued, then what will be the impacts upon the sparse life of these regions? The effects on offshore marine ecosystems are, by contrast, rather well known; the stirring up of great quantities of seabed silts usually has a positive effect in releasing nutrients into biological systems which are limited by lack of e.g. phosphorus and nitrogen. On the other hand, the silts may blanket and kill organisms and at the very least cause turbidity which inhibits photosynthesis. The institutional structure, too, is difficult to organise outside EEZs. One possibility is a UN Sea-bed Authority. But do they merely regulate the activity of mining companies or do they manage the resource on behalf of all the worlds' peoples?

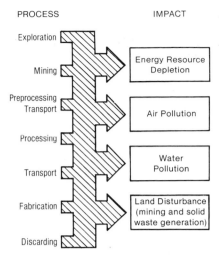

Fig. 2.20 Mineral uses and their impact upon components of the environment. Generally the lower the grade of ore, the stronger the environmental change wrought. (Compiled and redrawn from various sources).

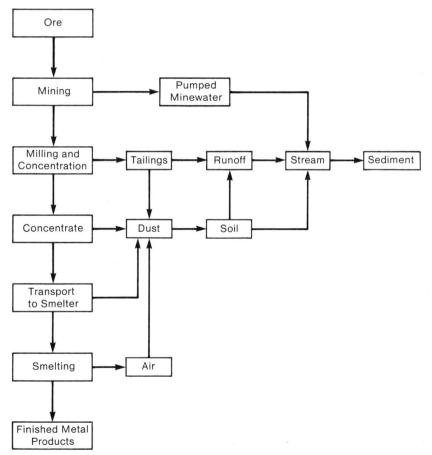

Fig. 2.21 The indirect transport of pollutants to the aquatic environment as a result of mining activities. M. Kelley, *Mining and the freshwater environment* London: Elsevier, 1988.

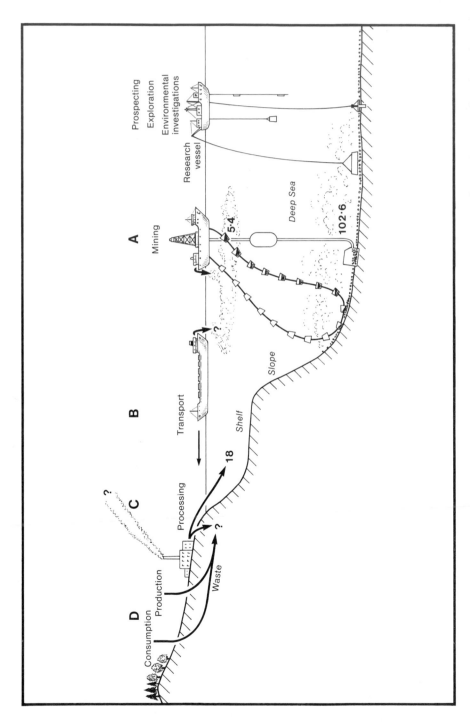

Fig. 2.22 A scenario for deep-sea production of manganese from nodules with onshore consumption. Contaminants produced are shown in million tonnes per year from a medium-sized operation of perhaps 10 vessels per year. (Redrawn from a number of sources).

It is no surprise, therefore, that in DCs there is usually strong government intervention in the minerals process to ensure some degree of amelioration of environmental impact; many LDC populations would like the same consideration extended to them but the strength of their governments vis à vis the need for foreign exchange and the power of MNCs often results in second-class treatment at best.

Mineral futures

Every so often, highly optimistic statements appear which point to the absolute amounts of minerals present in the crust of the earth and posit a non-problematic future for all concerned, i.e. everybody. Unless a very cheap and totally ubiquitous energy source becomes available, however, we are still in a context where concentrations are important, and likely to remain so. Therefore there is some element of scarcity and so all the forces of economics are brought into play. Together with government policies, this means that mineral futures are an outcome of such factors as the discovery of new deposits whether on land or at sea, the coming on stream of lower concentrations of minerals called forth by higher prices or improved technology, the discovery of different sources for certain products, e.g. sulphur as a by-product of power generation from coal and oil, the penetration of demands for recycling, the possibilities for substitution (could we have a carbon-based rather than an iron-based industrial civilisation?), and even the distant possibility of extra-terrestrial sources of minerals.

Given this array of variables, we need to look at all mineral forecasts with a degree of scepticism. The commonest forecasts are of two types: the first shows the downward trend of real prices of minerals and asserts that there is here evidence for oversupply rather than shortage. This argument needs to be viewed, of course, in the light of the terms of trade between suppliers and consumers and also in, for example, the employment policies of the mining companies, especially in LDCs where they may be paying very low wages. More pessimistic views comprise the second category (Fig 2.23), where known reserves are divided by actual and projected consumption rates and the life-expectancy in years of a particular mineral is extrapolated. This process is subject to all the variables outlined above and many think-tanks have made sophisticated models of supply, demand and technology: results of some of these are shown in Tables 2.24 and 2.25. Though in general the forecasts are more sunny than the simple reserves/use ratio, the lifetimes are never all that long, given the lead-times necessary in most complex industries for the development of new (as distinct from incrementally improved) technology. So impetus is given to re-use of e.g. metals not only in terms of ways of recovering used materials but designing products for (a) a long life in the first place and (b) easy recovery when the use-period is over.

Table 2.24 Life expectancies of selected mineral reserves late 1980s

	(1) Reserves	(2) Reserve base	(3) Addns to reserves 1950–74 (t)	(4) Ann Consmpt	(5) Expectancy: (3)/(4)
Lead (10^6t)	75,000	125,000	1.7×10^6	5,489	14 yr
Nickel (10^6t)	51,710	100,699	3.9×10^6	792	65 yr
Iron Ore (10^6t)	153,416	216,408	7.6×10^6	858	179 yr
Zinc (10^6t)	148	295	1.5×10^6	6	25 yr

Adapted from various sources

Note: This is a very crude calculation: expectancy may *increase* as new elements of the reserve base are tapped (Col. (3)); it may *decrease* if rates of use increase. So the numbers in col (4) are very provisional indicators only and depend on a number of assumptions with high inherent variability; see also Table 2.25

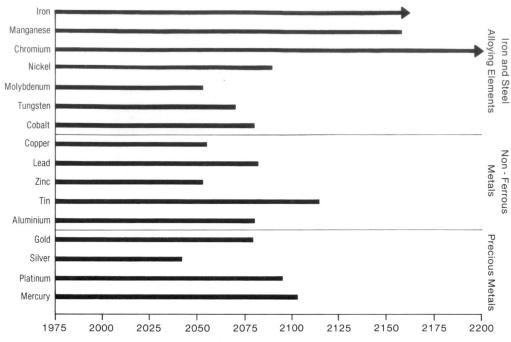

Fig. 2.23 Lifetimes of various mineral ores given certain assumptions about rates of use and ratio of reserves to resources. Thus such a diagram is only a very general indication of problems with future supplies. (Compiled from various sources).

Table 2.25   Variable life expectancies of selected minerals

	(0)	(1)	(2)	(3)	(4)
Lead	14	31	45	50	49
Nickel	65	47	56	73	92
Zinc	25	22	28	32	31

Sources: M. Holdgate et al (eds) *The world environment 1972–82*. Dublin: Tycooly Press, 1982, p. 195 (Cols (1)–(4) estimates by different authorities. Col (0) is col (5) from Table 2.24. Note how it is exceeded by most of the other estimates.

Futurology in this field has a long history (especially about fuel minerals) which has mostly turned out to be wrong. One key variable is substitution, which is considered in some detail by studies which suggest, for instance, that present patterns are likely to continue for 30–40 years. Thereafter the non-ferrous metals (alloy steels, aluminium, magnesium and titanium) would assume a greater importance. After the non-ferrous metals only virtually inexhaustible or recyclable materials can form the basis of a materials economy: wood, glass, plastics, iron and magnesium, are key features. A further problem is, as always, the sources and quantities of energy which will keep these materials moving

Table 2.26 Real price of selected minerals 1920–50

	1920	1940	1950
Coal	340	195	413
Copper	170	125	129
Iron	216	149	146
Lead	292	211	298
Zinc	301	281	335
Aluminium	647	297	217
Crude oil	547	205	278

Source: P. Dasgupta, Exhaustible resources, in L. Friday and R. Laskey (eds) *The fragile environment*. CUP, 1989, 107–26. The 'real price' uses the actual price relative to the price of capital of the time.

through the economic system. Costs of minerals, in general, seem to have followed a U-shaped curve this century (Table 2.26).

Since they are non-living, mineral resource systems ought to be simpler and hence easier to predict than those systems which depend upon the complex feedbacks of the organic world – not really so, in fact: the human-constructed fabric of economics, politics and technology can be every bit as dense and labyrinthine as nature's ecological webs and warps.

Non-renewables: summary

Concern over future supplies of all kinds of resources has sometimes invested the term 'non-renewable' with a kind of negative air, as if a normative judgement was being made that somehow it was wrong to use these materials. Nevertheless, current civilisations are built upon them in the most literal of senses, and the history of their use is a very long one indeed. At the same time, the growth of population, the rise of material expectations and the extra energy needed to extract poorer grades of minerals, all make a cornucopian future unlikely. Given that in most cases it is less energy-consumptive to recycle, and less costly overall to re-use then many societies will find it in the interests of the population (if not perhaps the shareholders of the mining companies) to move in the direction pointed out by their Green movements which have re-use as a first stage. Whether western societies have the capability to go to the next stage after that is another matter, for it involves thermodynamic and materials thrift brought about by longer lifetimes for many goods. Are we capable of adapting (except under the most pressing of conditions) to societies in which the lifetime of vehicles, for example, was 50 years or where a pair of shoes bought at age 17 would last the rest of one's life? Further, are all societies prepared to give up using non-renewable metals, other minerals, and energy in preparation for, and conduct of, warfare?

7

Environments as resources

The whole, not just the parts

In this chapter we are concerned with the totality of an environment as a resource for humans. There are a number of circumstances in which this takes place and at the outset some characteristics may be outlined:

- In most such systems, solar energy predominates over human-directed energy flows. In general, we deal with the less intensively managed and exploited ecosystems.

- Nevertheless, there is no *a priori* requirement that the environment has to be 'natural', i.e. in a pristine state of nature in which human societies have exerted no manipulative influences.

The kinds of environments, then, which come to have value in themselves to humans include for example assemblages of biological species together with their non-living environments, i.e. ecosystems *in toto*. These systems may be prized for a yield since one or more species can have a market price but the valued species is so embedded in the ecosystem that either (a) it cannot exist outside that system or (b) it has no worth if divorced from its customary habitat. An example of (a) might be a commercially marketable species of fungus which will grow only in beech woods on acid soils; and of (b) the value of a polar bear skin to a keen sport-hunter: the animal must have been shot in the wild, not as part of the culling programme of sick animals at the local zoo. Equally, such systems may be valued for non-consumptive pleasure: the attractions of the game parks of Africa are well known, as is the thrill of watching a rare species in its native setting, like for example the Indian tiger, or the gyr falcon in Iceland. Many nations make a good deal of money from tourism which is inspired by the wish of people to see wild animals and plants in natural-seeming and indigenous habitats. Then again, there is the value of such assemblages to science. The concern may be purely academic in the sense of, for instance, finding hitherto un-named species of plant and insect in tropical forests, or it may be instrumental. In this latter case, science may be interested in new species of plants because they contain base materials for new or improved drugs. Science may also perceive possible new food plants, or repositories of genetic material which it may be possible to breed into older domestic varieties or to splice into other species to confer upon them some desirable property.

Another major category of environments as integral units lies in the value of whole vistas as scenery. Here the biotic components of the ecosystems are still needed, as it were, but are less important than the appearance of the surface of the land (and often water) itself. The pleasures to be got from contemplating views, whether they are gentle or dramatic seem now to be well-engrained in the western worldview and the behaviour of tourists at certain places makes it clear that appreciation of scenery is expected of the visitor even though he or she may not really care very much for it. (I recall standing at the first overlook on the eastern approach to the Grand Canyon and being amazed at this sight which really lived up to all that had been said about it. A couple in a car with Florida plates drew up. Only the man got out, looked briefly at the Canyon and said, 'nothin' like this in Florida' and they drove off in the direction from which they had come.)

We should not get the impression that only large areas of very wild country fulfil the cultural requirements exemplified here. We can all think of examples of small areas of rural scenery or even an urban park which provides the necessary ingredients for peoples' pleasures. Thus it is that the garden is one of the prime examples of a heavily-manipulated ecosystem but nevertheless one which yields a great deal of pleasure to its users. It is generally characteristic of gardens that they yield useful products as well, with the balance varying from time to time and place to place: in the rural areas of developing countries the spices and medicinal plants have pride of place perhaps but in the West flowers and shade trees often predominate.

However, in a world which has undergone millenia of human impact and whose ecology is changing rapidly in some places, large areas of wild land have in many eyes a special place. Such very large tracts are usually termed 'wilderness'. They may have this character simply because they are areas marginal to economic use: unused land, as it were, like the interior of the Sahara or parts of the tundra of the USSR. Other parts have been recognised by nation-states as being so worthy of preservation in such a wild and unmanipulated condition that they enjoy special legal protection. The United States is a leader in this field, having 450 designated 'wilderness areas' totalling 360,000 km², but many nations have followed: even small and densely-populated countries like Japan. The world-scale example is Antarctica, which is governed by a special international treaty that dedicates the whole continent and surrounding seas to science and conservation.

It will be convenient, hence, to treat environments in two groups: the first of landscapes and ecosystems as resources, and then wilderness as a case of the value of a 'non-resource', though in many places there are pressures to convert such environments into resources of a conventional kind.

Ecosystems and landscapes as resources

As we learned in Part I, the impact of humans upon the natural environment has been a long and pervasive one, with the result that few ecosystems are pristine. The perception of human-wrought changes may be one of degradational tendencies or of creative alteration, though in either case the cultural valuation may not be consonant with the findings of ecological science. But any environment, altered or not, can be valued for its totality or at any rate for a particular element that is inseparable from the whole.

Ecosystems as resources

Given the centrality of food in human affairs ('Grub first, then ethics', said Berthold Brecht), the production of edible biota in near-natural and truly wild ecosystems must be of interest. If we look separately at non-industrial and industrial societies, then this is a useful place to remind ourselves that for those people without access to the village shop or the urban supermarket, wild foods may be an essential part of the diet. Such biota are usually species which are not amenable to domestication or which live in ecosystem themselves not 'tameable' because of size or ecology. Hence animals like river and lake fish, inshore marine mammals, honey, and plants like fungi, yams and nuts may all be important sources of flavour, variety and protein; nowhere more so perhaps than in the root and tuber subsistence economies of the tropics. Such ecologies sustain, too, the remaining hunter–gatherer economies of the world, where the wild foods have a more than ancillary importance in the diet. The actual dietary significance will vary according to the access to imported 'western' foodstuffs but their symbolic importance will be considerable. Thus in Alaska, the indigenous population has always hunted the bowhead whale. People could of course be fed upon hamburger meat from the Amazon basin but the hunt is an integral part of the culture. This has caused conflict with conservationists who see the bowhead populations depleted, like those of so many whales and want to impose quotas or even a moratorium, pointing out that though the image of the hunt is traditional, the technology used for it is highly modern. The seal plays a somewhat similar role in Inuit communities of the high Arctic of Canada, though happily it is not an endangered species. The Bushmen of the Kalahari who were studied in the 1960s showed a strong reliance upon one species of nut from a wild tree for most of their necessary energy and protein but the surrounding savanna was valued for the 50-odd species of edible animal which gave variety to the diet and the catching of which gave status to the men.

Many other subsistence or near-subsistence agricultural societies value wild country precisely as a source of additional foods. In the Amazon basin, the rivers may yield good catches of a variety of fish, just as the canopy of the forest may contain monkeys that are likewise sources of animal protein. In many places, modern technology like rifles has improved the success of the hunt though usually depleting the animal populations. Wild yams also are a useful source of food, from Australia to meso-America, and although it is possible to 'tame' some species of bee and keep them in hives near to a settlement, some others are best at a distance with the honey being taken at the appropriate season. For such people, too, some of the species of the wild are threatening and become non-resources in the sense that they are extirpated if possible: mammalian predators and snakes are examples.

In industrial societies, much wild food is a by-product of a primarily sporting activity but there are examples of species gathered from forests and other lands specifically as a seasonal food: near the small towns of Poland in September, for example, children gather large basket-fulls of fungi to sell. Similarly in an English autumn, heaths and moors are combed over by people gathering bilberries (*Vaccinium* spp) and pieces of scrubland and woodland edge are trodden nearly flat by the garnerers of blackberries (*Rubus* spp). Those into more alternative life-styles may gather many more species for foods, herbs or medicines and so visitors may be lucky enough to be offered nettle soup for lunch.

The killing of animals for pleasure is scarcely a new trait of human behaviour. All through history, access to animal populations for sport has been a sought-after prerogative of status, with careful control over the land resources and other paraphenalia: Louis X of France was so worried about the possibilities of rabies among his hunting dogs that he used to take them to Mass. In general the larger animals were reserved for the rich or the powerful. Only perhaps in North America is there a plenty of medium-sized mammals so that a form of democracy can prevail in that activity. The expense of such sports today can be very great but then so can that of many others: the hardware in the 'sportsman's' store is often equalled by that in the mountaineering shop or the ski boutique. In 1980 in the USA, 17 million hunters spent $5.6 bn, as well as giving $6 million as voluntary contributions to wildlife conservation organisations. Their opponents spent a mere $50 million in the promotion of their point of view. The pleasure of shooting and fishing can occasionally be buttressed by illegal sales of the bag, so that most valuable game animals are predated upon by poachers as well as the legal hunters.

Where wild country is concerned, then non-domesticated animals are usually taken as a controlled usufruct from the natural or near-natural ecosystems which have received a minimum of management. Thus remnant areas of African savanna outside game reserves and National Parks are still hunted over by safaris in which the rifle and not the camera are pointed at the animals, and the cull is restricted by what can be found rather than by management regulations, though voluntary exclusions, e.g. during breeding seasons and of gravid females, may be observed. The same would be true of bears in the taiga forests of the USSR. Species protection laws may be enacted however to protect animals which are migratory or which live outside reserves. There is, for example, an international treaty protecting the polar bear which however allows small quotas to indigenous populations, which they are free to use themselves or to employ as a source of monetary income from rich urban-based sportsmen. Sport fishing in rivers (for salmon for instance) is usually controlled by a mixture of expensive licences and catch limits and there is often conflict between rod fishermen along the river and commercial net operators in estuaries and off-shore.

Occasionally wild terrain is managed intensively for production of quarry for sport, as in the case of the grouse moors of England and Scotland. Here a high density of the main food (heather – *Calluna vulgaris*) of the grouse is maintained by rotational burning, and other inputs such as the provision of grit, predator control, and the exclusion of disturbance during nesting, all contribute to high densities of grouse on the moors at the opening of the season on the 'glorious twelfth' of August.

Tamer ecosystems like agricultural lands, small relict woodlands and enclosed ponds and lakes can also yield their share of sport animals: stocking of lakes, for instance, is common: this may be with trout and have a catch limit or with coarse fish which are usually returned to the water at the end of the day. Small lakes in peri-urban parks or large ex-gravel pits may get this treatment. Birds like the pheasant may be reared under domestic hens and hand-fed, protected from foxes and hawks and then released into woods from which beaters expel them in numbers large enough to provide sportsmen with a considerable number of targets in a short time. However the prime example is the hunting of foxes in lowland England and in Virginia. The maintenance of rights of way, the type of fencing, the conservation of small copses all form part of the management for this sport which continues alongside the normal agricultural use of the land; other resources involved are horses for which grassland is needed, and hounds for which kennels well away from light sleepers are desirable. In southern Europe particularly, farmland and orchards are the same as anywhere else when it comes to the indiscriminate shooting and netting of any species of bird foolish enough to move on the day that the hunting season opens in the autumn. The mortality rate among cattle and humans rises at the same time.

Attempts have been made, especially in Africa, to perpetuate wild species of ungulate by ranching them. Meat yields are high and the subsequent environmental impact lower than with domestic animals like cattle. Though useful ecologically, it does not seem to have proved culturally attractive to Africans.

Since ecosystems are by definition composed of living species, the very existence and diversity of these species is itself a resource. These entities may be potentially of use to human societies, their genetic material may also be valuable for breeding into existing domesticates, and they are often of intrinsic value to science. All these are threatened by the accelerated rates of extinction being brought about by habitat change, pollution and over-exploitation. It is, of course, true that extinction is the normal fate of a species: probably 90% of all species that have ever lived are now extinct, but human societies seem to be intent on accelerating the 'natural' rate. Estimates of the total of biological species present vary, with a number of the order of 3–10 million being acceptable to most authorities (and 1–5 million of them having been described by science) but suggestions of 30 million insect species in tropical forests alone are also current. The present rate of extinction places perhaps 1000 species of birds and mammals into the endangered category, along with very many species of lower animals and plants. In the USA alone in the 1980s, 74 species of plant were considered as endangered or threatened. It is probably facile to extrapolate current rates into the future, though to do so produces the frightening figure of 25% of the world's species being lost (and as the slogan says, 'extinction is for ever') in the next 3 or 4 decades. But even accepting lower rates of loss as the problem is realised, we need to remember that if 100 km^2 of deciduous forest in the temperate zones is replaced by a plantation crop of a vegetatively propagated species, then the genetic variety of the human-directed ecosystem is about 2% of the original forest. If we add the losses from accompanying species, then the crop will be a few tenths of one per cent of the original. The loss of species is not of course even: particular areas of concern are the tropical forests and islands, and taxa focused upon are plants which may have useful properties, endemic species of very restricted ranges, and scarce predators at the tops of long food chains. Thus in many places it is very specialised species which are becoming extinct and their place given over to generalists, rather analogous to Man itself as the great generalist species.

The role of species diversity in ecosystems has been much debated and it seems probable that ecosystem stability is not necessarily dependent upon a high diversity of species. Nevertheless, disruption of the patterns of evolution in ecological communities is sometimes likely to be disastrous and nowhere more so than in the loss of genetic material with potential for further evolutionary transformation.

Emphasis is often given in discussion to wild species. However, there is an enormous gene pool in the indigenous varieties of many cultivated crops. Modern crop plants tend to be genetically very uniform since this pre-adapts them to the requirements of a machine age in terms of response to fertilisers and irrigation, harvesting time and processing qualities. Thus Greece has lost 95% of its wheat varieties in 40 years, and in Europe generally only 30 breeds of cattle are present in viable numbers, with 120 likely to disappear. The whole US soybean crop is derived from just six plants imported from Asia. In 1970 the maize crop of the USA was reduced to 50% of its normal level by an infestation of leaf blight. Most varieties of corn grown in the US shared a single gene that made them susceptible to the pest. The presence of other varieties made possible a new strain resistant to that disease. So protecting the genetic resource in many crop plants is now very important especially since desirable categories in these older varieties can now be 'spliced' directly into commercial strains as well as bred in by the older methods in the greenhouse and the experimental plot. Loss of a variety or species now becomes not merely the loss of a book from a library but the loss of the pages from a loose-leaf volume which could be photocopied and inserted into other volumes so as to make them more useful. This endeavour, too, has its martyrs. In Peru, Sendero Luminoso guerillas murdered the man whose job it was to protect the world collection of potatoes.

So what other resources should be invested in conserving species variety and the load of genetic information which is contained therein? The two methods most discussed are called *in situ* and *ex situ* conservation. The first requires the designation and management of reserves of various kinds in which the total ecosystem is protected and is mainly applicable

Plate 2.22 An area devoted primarily to the protection of wild animals: in this case the relatively rare rhinoceros. Here in Kenya, there is a considerable income of foreign currency from tourists wishing to see such species, usually in relative comfort and safety.

to wild species. The second is the storage of essential genetic material in zoological and botanical gardens and as banks of seeds and sperm; eventually ova will probably be added. The classic protection devices are the nature reserves and national parks of the world, although the latter may sometimes be oriented as much towards tourism as to species protection (Plate 2.22). One estimate suggests that 1.3×10^9 ha need to be managed thus to enclose a sample of each of the earth's major biogeographic provinces. In the late 1980s there were about 425×10^6 ha so designated but their distribution focuses upon Boreal forests and tundras of the northern hemisphere. Hence areas like tropical forests, grasslands and the Mediterranean are badly represented. Overall, there are nearly 200 biogeographic provinces in the world but one in eight is not presented by a single park or reserve.

The best type of reserve is the Biosphere Reserve, a concept of UNESCO's Man and the Biosphere (MAB) programme (Fig 2.24). These reserves are of a size sufficient to contain whole ecosystems within their natural boundaries, have buffer zones against outside influences and generally carry out research and educational functions as well. In the late 1980s there were some 260 of these in nearly 70 nations. If one biome has to be singled out for conservation, it is the tropical moist forests or TMFs. Apart from all the other good reasons (pp 82–5) for their protection, they contain the greatest species diversity in the world and a great many useful materials have had their origin in these plants. It seems unlikely that they will yield future food plants of significance, but there seems likely to be a great storehouse of waxes, oils and alkaloids. The rosy periwinkle, for example, is a Madagascar TMF plant which has been processed to give two anti-cancer drugs. It is estimated that 25% of prescription drugs have their origins in TMF plants.

Ex situ conservation recognises that a species will not survive outside carefully controlled conditions. For wild species, it has been notably employed in zoos and botanical gardens

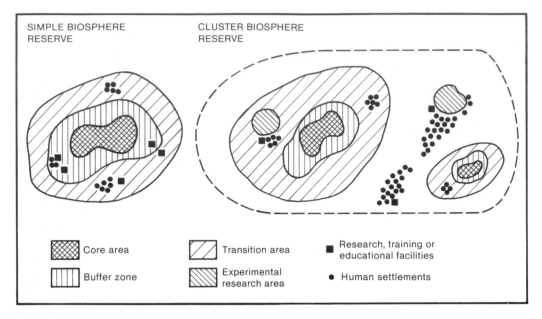

Fig. 2.24 Two types of layout of a biosphere reserve: a simple one and a more complex cluster. Where there is land and political will, the second is much preferable. *World resources 1988–89.* NY: Basic Books Inc, 1988.

(Plate 2.23), sometimes to the point where enough breeding success has been achieved to release specimens back into the wild. However, its main thrust at present is in the collection of plant material in gene banks, usually in the form of seeds stored in conditions of low temperature and humidity. Thus the seeds of a rare (only 2000 plants in all) perennial maize discovered in 1978 in Mexico can be held against the day when they can be bred into the common annual maizes, with the prospect of perennial maize production as well as resistance to some common diseases. The case of rice is instructive: one germplasm bank (at the IRRI in the Phillipines) contains over 70,000 Asian cultivars and 2000 wild varieties, although it is known that China alone grows about 40,000 different varieties of rice. Yet this form of conservation has its vulnerabilities: not all seeds can be stored without deterioration and even those seeds which seem viable after long periods of dormancy in cool and dry drawers must deteriorate one day; a prolonged power cut might destroy a large proportion of the bank entirely. A last objection is that this method of conservation literally freezes evolution and that the immense natural potential of a variety to continue to adapt to all kinds of changing environmental factors is stopped in its tracks. Thus areas of land (which is to say the diversion of another resource) may need to be set aside as, so to speak, 'gene parks' for a better solution to problems of the loss of genetic variety. Just as farmers in the West are often paid not to grow certain crops, so maybe farmers and foresters everywhere need to be paid to conserve indigenous varieties and their sustaining ecosystems.

In terms of the loss of species, there is no doubt that time is critical. Once a species is lost, it cannot be reconstructed and perhaps no other field covered by this book falls more neatly under John Locke's aphorism of the 17th century that 'hell is truth seen too late'. One of the main problems, as might be expected, is where to start once the world's 30 main crop plants (which account for 95% of food production) have been covered. At present, the main objects of concern are the wild legumes and the Gramineae, the grass family.

Plate 2.23 Zoos have a valuable function for *ex-situ* conservation of rare and threatened species. This is a newly-born orang-utan at the Chicago Zoo. The role of zoos as a form of public entertainment is more questioned, from both the ethical viewpoint and that of commercial viability.

The preservation of species and genetic variety is mostly seen in practical, even commercial, terms. However it is not seriously in dispute that many people get a great deal of pleasure from ecosystems experienced as wholes. Experience in this case means mostly the response to the visual qualities of the environment and its biota, but other senses may also play a part: the smell of the African savannas in the heat or of deciduous woodland, the sound of lions roaring in the night or of a nightingale singing, are simple examples. The easy proof of the attractions of wildlife for pleasure is the number of packaged (and usually expensive) holidays that are on offer in the West which centre about wildlife in a 'natural' setting. To some extent, of course, these overlap with the idea of scenery as a resource (see below) but here the clientèle is likely to be different: a more specialist and more committed type of person. In fact wildlife holidays are the equivalent of, for example, art and architecture tours though the certainty of seeing Stoddart's peccary is possibly less than Giorgione's *The Tempest*, even given the uncertainty of the opening hours of galleries and museums in Italy.

The target of observation for wildlife enthusiasts, whether packaged or not, is first of all likely to be wild places in which there is a reasonable density of plants or animals. Some are drawn by the whole ecosystem and its relationships, though more wish to see, e.g. alpine flowers or seabirds. Iceland might be a good example of the whole being attractive to a limited number of visitors who perceive in the inter-relationships of vulcanism, glaciation, northerly latitude and the sea a set of ecosystems supporting many species of plant and animal which have been harried out of more smiling countries. Thus Iceland is for them a kind of total refuge with rewards of sightings which outweigh the low cloud, cold rain and highly active insects.

Taken further, this type of demand is also made upon Antarctica. Here, the principal mode of access is the cruise ship, so that the extremities of that continent are experienced in more comfort than that of a pony-trek in Iceland. Here again, the combination of a lack of obvious human influence, vulcanism and glaciation, the sea and large bird (and mammals in the case of the southern continent) populations are the attraction, with some plants added on the outer rim of islands, though the Kerguelen cabbage is not perhaps the most obvious feature of the advertisements.

Possibly more common, though, is the attraction of a special and world-famous assemblage of animals, and round those of central and East Africa has been built a lucrative tourist industry which is very important in the economies of e.g. Kenya, Tanzania and Zambia. The hunting safari of colonial days has now been replaced by the picture-taking outing conducted from special hides or the open top of a minibus. Only the most élite actually use their feet. The equivalent of the cruise ship is moored in the National Park interior and nocturnal fauna can often be viewed from a verandah giving access on the one side to a flood-lit pool and salt-licks and the other to gin and tonic, with ice. The yield is a good profit to the tour operators, some income to the host country, some employment, and to the visitors a sense of nearness to something primeval, perhaps, plus enough pictures to bore the neighbours and the WI for several months.

Lastly, there are those who concentrate on special groups, of whom the bird-watchers are the most prolific. Like other specialists they can be oblivious to other parts of the ecosystem and have been known to damage habitats severely in order to get the desired view or picture. Indeed, some seem less interested in the bird as a living organism than as a tick in the year-list or life-list and many a rarity has been hounded to death by the activity of twitchers who allow it no rest or time to feed. In the UK, for example, there is a telephone line which gives the locations of recently arrived rare species of bird.

If we were to measure pleasure, however, the greatest aggregate of it in the case of birds would probably come from sightings of just-slightly-unusual species in gardens, which gives us the clue to an ecosystem which neatly rounds off this section on whole systems as resources. For the garden is typically both for pleasure and utility: it gives as it were both fruit and flowers. Not every garden now or in history has done so, but there is a strong element of both in most examples. In hard-pressed developing nations, of course, a yard area attached to a dwelling will if possible be productive (not of staples but more likely of spices, herbs and medicinal plants) but may well include one or more large trees whose shade is a source of relief to the inhabitants.

The history of gardens reveals a deep-seated dichotomy between the large and the small, with the former belonging to aristocracies and their more democratic equivalents. The latter are now often found attached to homes in the west, especially to those not built in cities in the 19th century. A typical layout will confirm that although an expanse of grass and flower beds are probably central to the design, an area of vegetable and herb production and possibly also a fruit tree are also likely to be found. Where the plot is too small for all of these, then it may be possible to rent a leisure-garden or allotment on the edge of a town in order to grow fruit and vegetables. In the case of houses without any adjoining land, then window-boxes are brought into service, or any patch of gravel or concrete in the hands of the Japanese can acquire remarkable aesthetic qualities. In the west at least, gardens share in the prevailing intensities of energy and materials uses. The quantity of energy expended on a suburban lawn, for example, is easily equivalent to that lavished per unit area on a field of wheat, which is after all mown only once in the year; watering the grass puts up the energy cost as well. In terms of output, the intensity of labour and materials means that the productivity of edible materials is far higher than in conventional agriculture: to raise the production of vegetables, the best land use policy is probably to build semi-detached houses.

In the past, the pleasures of hunting were, for the rich, ensured by the enclosure of parks which kept the animals in and the unwashed out. An 18th century version was the landscape garden designed to illustrate philosophical and artistic principles rather than organised bloodsports. The descendants of both have in many places passed into public ownership and are now for general enjoyment, though the usefulness aspect occasionally survives if the municipal nurseries are found in a corner behind a high wall. Private owners of large houses and wide spreading landscape parks are likely to raise income by providing attractions beyond those simply of avenues of trees and neo-classical temples of Flora. The safari park is one instance, with visitors paying to drive their cars through paddocks vegetated with a savanna of deciduous oaks and elms and populated with baboons, giraffes and lions. Pleasure and profit neatly combine, in central England as in central Africa.

Gardens are also characteristic of communities (Plate 2.24). A monastery or convent today is as likely to prize its garden as much as its medieval precursor. Production is

Plate 2.24 The garden represents a more or less complete transformation of nature. It can often therefore be installed far from its native location. This Japanese garden was constructed in Gateshead (England) during a Garden Festival in 1990, which saw the flowering of a local gasholder as well.

all-important here, as is also the pleasure-giving qualities of the garden, although the language may be different and the Dark Ages excuse that the flowers and the shade are really for the elderly and the ill may also be exercised slightly defensively. But a cathedral close will unashamedly flaunt its roses, and a college quadrangle its gillyflowers and hollyhocks. Indeed, the monastic cloister with its enclosure and central fountain or tree is the obvious precursor of the educational precinct. This symbolisation of a cosmic order is less apparent in modern university and polytechnic buildings: their iconography is more likely to derive from the imposed poverty of a suspicious and authoritarian government or the trendiness of the PR company employed by an industrial sponsor.

Thus the garden has always been a potent cultural indicator. Because of the intensity of effort owners are willing to expend, it can always be made and remade in an image conceived in somebody's head. Antarctica we have to take or leave (more or less) but the back yard can reflect us as we are or even as we would like to be.

Landscapes as resources

The psychological basis of individual reaction to surroundings is very complex and has engendered a large and highly specialised literature. It is probably impossible to distinguish between a genuine existential encounter between a person and, say, a sunset or a glimpse of a kingfisher, and a more programmed interaction in which we react to these things because we have been trained to do so by education, television or paintings. (We might reflect that in the 18th century landscapes were called 'picturesque' because they resembled pictures.) That the aesthetics of our everyday surroundings are an important element of choice (for those who have choice) is obvious: see for example the ways in which housebuilders and estate agent advertise their goods, or the settings in which TV advertisements place expensive items like cars.

One way in which these reactions manifest themselves is in outdoor recreation. Some of this is urban-based but a great deal is either in rural surroundings or in imitations thereof like artifical ski slopes or former railway tracks which are now walkways and bridle paths. Just as food is produced with the aid of a huge spectrum of technologies, so the landscape resources of an environment are tackled with a great range of technologies. At the simple end of the spectrum is a pair of stout shoes (or, today, old trainers) for walking through undemanding terrain. This may escalate to boots, high-tech clothing, and survival bags for the real mountains, to even higher-tech gear for skiing (Plate 2.25) accompanied by large quantities of *in situ* technology such as chair-lifts and gondola-cars (Fig 2.25). Most of these are made possible by combustion-powered vehicles and indeed driving cars (and winter vehicles like snowmobiles) itself ranks as one of the most popular outdoor recreations. There is also some reaction to the involvement of so much technology, so purists take up back-packing in wilderness areas and nude bathing.

Most developed nations have prepared reports on the habits and preferences of their own populations for outdoor recreation and undertaken surveys of what are usually called 'recreation resources', i.e. those land and water areas which can be devoted to these uses either as part of public provision or as a commodity sold by private enterprise. In many cases these surveys are also designed to catalogue the development potential for tourism, i.e. the attraction of visitors from outside the region or country (see later sections). The types of environment favoured for recreation depend on many factors which relate to age, income and time availability of the recreationists as well as the obvious 'natural' features. These might include the presence of water, a certain wildness (but not too much for most people), the presence of some wild life (providing it does not threaten), and a view of high lands and mountains, though preferably accompanied by the means to scale the heights

Plate 2.25 In the French Alps, skiing is responsible for the import of large quantities of technology to high altitudes. The isolation of this tow-head emphasises the distances involved. In other places, considerable developments of cafés, hotels and shops may be perched on top of mountains.

mechanically and find a café at the top. Such features are common to most of the western world and hence find their way into many developing countries which aspire to a tourist industry.

An important element in this type of recreation is that people should perceive that they are in the midst of 'natural environment'. They rarely are, of course, but the common equation of 'wild' with 'natural' enables them to bridge that cognitive gap with some ease. Hence users of well-developed trailer parks (with electricity hook-ups, hot showers and coke machines) have been known to fill in questionnaires which allowed them to assert that they were having a wilderness experience. Different social groups are apt to display different tolerances and preferences, too. In England, there is an anathema to recent coniferous plantations on the part of the recreational *cognoscenti* and their organisations. Yet the untutored visit them in large numbers and show very little sign of finding them unattractive.

But the obviously human-constructed is not necessarily rejected for it can be landscaped to look less 'artificial' as with the gravel pits used for sailing and fishing discussed in Chapter 1, or indeed accepted for itself: the open-air museum of rural or industrial bygones is a recreational environment (with some educational purpose as well) which everybody knows to have been assembled on an available site: it is not 'real' but seems to lose little attraction thereby. Likewise, the urban park is scarcely a natural environment but it is sufficiently different from its context to be attractive to those with little money to travel or with only a little time at their disposal. (There is a school of thought that says city parks physiognomically mimic savannas and are therefore popular because they hark back to the

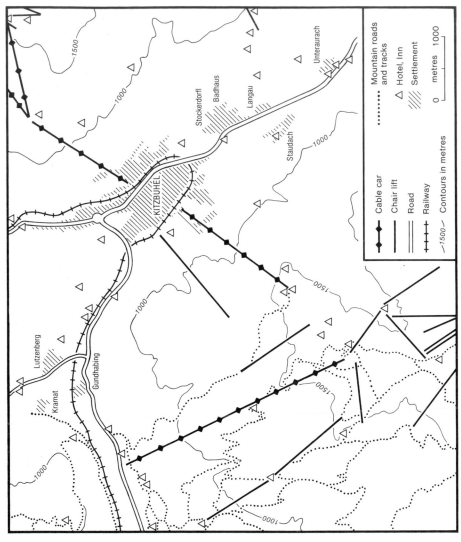

Fig. 2.25   The intrusion of technology into an Alpine landscape near Kitzbühel in Austria for skiing: lifts and roads.

environment in which the genus *Homo* evolved, but it need not detain us long.)

Outdoor recreation exerts many ecological impacts upon the resource which it employs. Nearly all aspects of a local ecosystem are prone to change under the influence of construction, of human presence, of human feet in large numbers, of air contamination from vehicle exhausts, of the disposal problems of the sewage and solid wastes of visitors, and in many other ways both subtle and obvious. At an obvious level, a pathway through the countryside exhibits a zonation of bare area flanked by species which are successively able to tolerate trampling. Heavy use spreads the path laterally, especially in wet patches and the result is something of a combination of a six-lane freeway and gulley erosion. Large numbers of humans change the feeding habits of biota, and many animal species have adapted to a modified role as scavengers in car parks and picnic sites; brown bears in North America are one much-encountered example. Large-scale impacts are evident in the case of some winter sport areas. Here, there is considerable visual intrusion, though this is usually accepted even though the tops of some mountains in e.g. Switzerland and Austria bear an increasing resemblance to a main line railway terminus. The maintenance and use of the *pistes* packs down the snow, reducing its insulation cover to the soil underneath. This snow may become more prone to avalanches but more significantly the soil underneath bears a higher risk of becoming a mud-slide after the snow has melted. The role of skiing in preventing the growth of young trees only makes this problem worse. The desire to get to remote places make some fragile environments very vulnerable to rapid change: trekkers in Nepal are held to have been instrumental in further depleting forests in the Himalaya, for instance.

Landscapes as resources seem to exhibit a particular plasticity. While each culture may have a special scale of preferences, many people within that society accept modifications of it readily, generally preferring the activity over the 'correctness' of the environment in which it takes place. Always, though, there is likely to be a group of conservative 'purists'. Most affected groups however are willing to adapt their native preferences if there is money to be made from attracting outsiders: if all else fails economically then try tourism. To attract and retain tourists some superficial cleanups may take place: the Li river of the Guilin region of China is one example, since the limestone pinnacles and the trained cormorants are best seen from river boats. So pollution control and bank erosion control have been enforced.

It is fashionable to decry the adage that 'travel broadens the mind' but that does not prevent it from becoming a very popular activity, with 60–70% of all international arrivals being for holidays. Since the total of such arrivals in the 1980s ran at about 300×10^6 per year, this is no mean economic activity. In the OECD countries in the same period, travel of all kinds accounted for an average of 0.9% of GDP, with Austria highest at 7.7% and Japan lowest at 0.1%. In selected developing nations as well, the tourist industry is very important, with Kenya (relying on game parks and coastline) as the outstanding example. Here, there are some 700,000 tourists per year, worth £200 million per year.

The key to the resources needed for tourism may perhaps be approached by categorising travel motivations into two divisions: sunlust and wanderlust. The former is at heart a 'push' factor, in the sense that the traveller above all wants to be somewhere different where she or he can 'relax' and demands the most pleasant of sourroundings in which to do so. Clearly, sand, sea and sun are the basis of this type of attitude, often together with the opportunity to add to the lust inventory and doubtless the AIDS statistics. But if the 'essentials' are present, then the rest of the environment is of lesser consequence: accommodation, sanitation, food and entertainment need not be different (and indeed often better had not be) from home. Hence the appearance of an American hamburger chain not far from Mozart's birthplace. Wanderlust, however, is different since the rewards are in terms of personal development: getting your bronze badge for scaling 80 points-worth of Austrian peaks; hearing Vivaldi in the church for which he composed, reflecting on the symbolic role of the sacred crane in Hokkaido, are more the stuff of wanderlust and no matter if flour dumplings, dryish pasta and glutinous rice appear on the menu with a predictable frequency. Clearly, sunlust and wanderlust are not always separable (they tend to meet in the bathroom for instance) but they provide a guide to resource demands.

Sunlust is *par excellence* a coastal phenomenon and as such its clientèle has made for those seas which offer reliable sunshine for at least part of the year, beaches which if not present can be constructed and maintained relatively cheaply, and the presence of the sea or a very large lake. The water quality need not be very high if swimming pools are available. Into such a place the accommodation is inserted, often at a high density, and producing urbanisation even if it did not pre-exist. Examples of this phenomenon are seen world-wide, though perhaps Spain and the Caribbean hold the rank of market leaders. Developing countries like Kenya, The Seychelles and The Gambia are also in this business. The pre-eminence of sunlust is perhaps being challenged however by its off-season version of snowlust, with that form of water replacing the sea. Because of high investment costs, this tends to be confined to richer parts of the world, like the Alpine parts of Europe and Japan, cordilleran North America, and Scandinavia. Ski resorts are hard to find in Tibet and Algeria.

The environmental impacts of sun/snow-lust development are well understood and indeed have been mentioned already. The rapid urbanisation associated with sea-side development usually places demands upon an inadequate infrastructure, so that airports, roads, medical facilities and waste disposal are all likely to be under great strain. Holiday-makers may find uncompleted hotels, early-morning building work and sewage in the sea all unacceptable and they are leading to a return to cleaner areas: Germans for example forsaking the Costa del Sol in favour of the Chiemsee in Bavaria very largely on environmental grounds. Winter sports enthusiasts do not in general have to see the environmental results of their enjoyment and so are less likely to want to seek alternatives. Are, then, the seekers after far-away places low-impact tourists who fit, camouflage-like, into their temporarily adopted landscapes? There are indeed many who try to be in that category, who take themselves or their cars or bikes 'off the beaten track' and who try to live off and through the land, whether it be country or town. They demand no special facilities and indeed are disappointed if such exist since they want to see the authentic place, warts and all. Middle class dinner parties tend to have at least one couple who have found a bit of the 'real Spain', or got right off the tourist track in this 'super little place outside Split'. What they have so often come across is of course a genuine piece of poverty.

Before long, however, the low-impact wanderlusters are bound to want to see one or other of the great cultural sights of the world: they are that kind of people. Here they join with packaged wanderlusters and many of the sunlusters having a day off from the tanning grill. So Venice in summer is host not only to devotees of Bellini and the sight of the Lagoon at sunrise but to coaches from Austrian villages that have set off at 0500 and will return their passengers to their beds at 0200 the next day. And presumably sunlusters on the Italian Riviera are offered a day in the land of Mozart and the Sound of Music. Where urban cultural resources are concerned we might imagine that environmental impact of the kind normally described for tourism does not exist: we are in a pre-existing city after all. Such a picture is not entirely true, for the extra people are likely to add to any strains on the infrastructure and if they are present for only a short season then it may not be deemed worth investing any more money to cope with them. High densities of tourists always have social effects (which are a bit beyond our task here) but apart from the growth of souvenir shops and expensive cafés one salient environmental effect is noise. Although banned from the interiors of many buildings, especially churches, parties being addressed by a guide are very intrusive to others' experience; this is the more so if, as now happens in some European cities and Japanese cultural sites (and maybe elsewhere: readers are invited to send me examples), it takes two guides to address a party of visitors: one to speak through the megaphone and another to carry the battery pack that powers it.

Eventually the pressures are such that radical solutions are canvassed. The case of Stonehenge, the ruins of a prehistoric stone circle of unknown but presumably ritual purpose in southern England, is instructive. The pressure of visitors meant that they had to be routed away from the stones and to see them only from a respectful distance: to get in amongst them and look outwards or to touch them is denied. Inevitably the suggestion has been made that most visitors would be content with a replica. Similar settings abound nearby and modern plastics could replicate the stones very well; even better, the 'original' Stonehenge in one of the many imaginative reconstructions of its Neolithic state could be rebuilt. It sounds absurd to some ears but think of the popularity of the ultimate in fibre-glass replicas: Disneyland and its offshoots and imitators. If the cultural experience you want does not exist then (as with gardens) build it and the folks will pay in their millions to experience an imagined view of the past, or of a technology-laden future or, even better, of a complete fantasy: not for nothing is the prototype dominated by a fairy castle. *Future Worlds* or its precursor *Westworld* are the movies to see. Such concepts can be grafted on to the more mundane: the excavations of Viking York in northern England are displayed as a reconstructed town through which visitors pass in electric trolleys and are assailed by the smells as well as the sights and sounds; only in February is the queue less than an hour long, apparently.

These tourist experiences point us in a couple of directions. The first is that of the general relations of resource use and the biosphere, where it may increasingly be necessary to decouple the economic processes from the biophysical ones. The second is that tourism may be wrongly classified, as we have done here, with the renewable resource uses. Above some kind of threshold level (not easily determined, it must be said) tourism destroys the resources upon which it was originally based. The experiences at any rate of the wanderlust tourist become progressively detached from authentic contact from a history or a culture and made into something *ersatz*. At the same time, pressure of numbers destroys the foundations of any experience (and the habitat of the indigenes) in an awful analogy with the pressure of a fast-growing population upon food or fuelwood resources. But it is, these days, a brave host who turns away guests rather than building an extension for them.

Wilderness

How much of the planet's surface is still dominated by the flows of nature, where solar energy is still the driving force and where human activity can be said to be insignificant? Such land is usually labelled 'wilderness', which is not a term applied to the sea which nevertheless comprises by far the greatest area of near-pristine ecosystems. On land, an inventory compiled in the 1980s was made from the US Jet and Operational Navigation Charts at scales of 1:1M and 1:2M. To be called wilderness, an area had to be 400,000 ha (1 M acres) in extent and be free from evidence of roads, settlements, buildings, transportation facilities, resource developments and constructions such as pipelines and logging camps. Thus it is land which is without permanent human settlement, not regularly cultivated nor heavily and continuously grazed. But much of it will have been lightly used (e.g. by pastoralists): few areas of completely pristine land exist. The result (Fig 2.26)

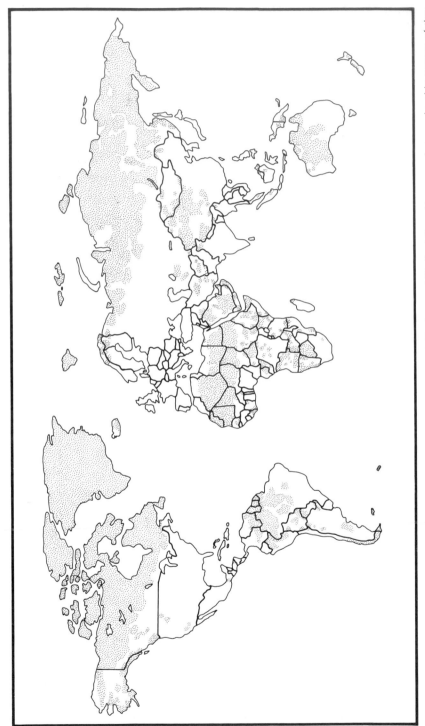

Fig. 2.26　The remaining wilderness in the world, 1980s. Source: J. M. McCloskey and H. Spalding, 'A reconnaissance-level inventory of the amount of wilderness remaining in the world' *Ambio* 18 (4), 221–7 1989.

suggests that there are 48×10^6 ha of such lands in 1039 tracts, nearly half of which is polar, and only 8-6% of these areas have some form of protected status.

'Unused' lands

The distribution of all kinds of wilderness land by continent and by biome is shown in Table 2.27. The dominance of Antarctica has been recognised in its special legal status (see below), thereafter Greenland, Canada, Australia and the USSR have most to contribute, with China, Brazil and Algeria also being important. Europe is so made-over and settled that it has few major wildernesses and indeed 108 countries in total had no wilderness tracts under this definition. By biome, tundras and deserts gain the most representation, with the boreal forests and the TMFs following. Many types of wild system have few remnants of the size to be recognised in this inventory, including many wetlands and island ecosystems. This encompasses the generally accepted idea that wilderness areas should have a minimum size in order that the ecosystems should be able to function in a natural manner.

Outside legally protected areas, wilderness areas are marginal in both settlement and

Table 2.27 Wilderness distribution by continent and by biome, 1980s

Continent	Wilderness area km $\times$ 10^6	%	No of areas
Antarctica	13.2	100	1
N. America	9.0	37.5	85
Africa	8.2	27.5	434
USSR	7.5	33.6	182
S. America	3.7	20.8	90
Australia	2.3	27.9	91
Europe	0.1	2.8	11
WORLD	48.0	32.3	1039
Biome			
Tundra	20.0	41.7	100
Deserts	9.3	19.4	389
Boreal coniferous	8.8	18.3	120
TMF's	3.0	6.3	77
Mountains	1.9	4.1	76
Cold deserts	1.4	3.1	51
Dry tropical forests	1.4	3.0	120
Tropical & savanna	0.7	1.5	33
Temperate rainforest	0.4	0.9	35
Temperate broadleafs	0.3	0.6	20
Temperate grasslands	0.2	0.6	24
Evergreen sclerophyll forests	0.1	0.4	7
Islands	.09	0.2	7
WORLD	48.0	100	1039

N.B. Figures in Col. 1 are rounded, hence occasional differences in resultant numbers in Col. 2

Source: J. M. McCloskey and Heather Spalding; 'A reconnaissance-level inventory of the amount of wilderness remaining in the World, *Ambio* 18, 1989, 221–7.

economic terms. That is to say, they are either so remote that it is too expensive at present to recover any resources they may possess, or that their environment is so hostile as to prevent them being perceived as resources at all. Neither of these situations are immutable and so any calculations of wilderness areas need to have a date attached to them. But marginality does not necessarily imply the absence of humans and so some of the areas denoted in Fig 2.26 inevitably have some human uses of a conventional kind. Some of these allow inclusion in the definition paraphrased above since they sit so lightly on the land: for example, the use of the high deserts of central Asia by nomadic pastoralists, or the occupance of the high tundra by small populations of Inuit. The present methodology does not allow the exclusion of off-road vehicle tracks, since that would require either the addition of remotely sensed data or ground verification. Parts of Australia are known to be heavily used in this way. On the other hand, where heavy grazing has meant the encroachment of sage-brush in the western USA, then such areas have been excluded. Some areas, too, have single isolated points of human activity which, if supplied by air, may have little impact upon the ecosystems outside their immediate locality: oil rigs in deserts and radar stations in the tundra are obvious examples.

The futures of such marginal lands are difficult to predict. They may stay more or less exactly as they are with, as it were, nobody taking any notice of them. But there may be a move into them in order to exploit resources such as minerals or they may be transected by channels of communications (if a new road were built for instance) or other linear features such as pipelines. Yet again, a changed set of attitudes may mean that some will move into an officially protected status.

Designated wildernesses

In the 1960s it became generally accepted in the West that not all parts of the earth should be developed and that there were good reasons for making sure that wilderness areas persisted. In part, the motivation was instrumental in the sense that they contained biological species for example, which might in future be useful to human societies; there was also the less easily articulated feeling that the wild had an important role to play in human psychology and there has been added the viewpoint that other species have a 'right' to exist in their natural state. Thus, led by the USA, several nations have designated wilderness areas. Sometimes these have a special label but they may also be called national parks, landscape protection areas or game reserves, depending upon national tradition. The United States' Wilderness Act of 1964 is especially interesting since it essays an operational definition of wilderness as land which 'generally appears to have been affected primarily by the forces of nature, with the imprint of man's works substantially unnoticeable'. The key word in some ways is 'substantially': there is no requirement that the land be in an absolutely pristine condition nor that it must always have been so.

In spite of the moral force of the arguments proposed and the impetus given by the campaigns waged by several international conservation bodies, only 8.6% of the wilderness of the world (4.1 out of 48.0×10^6 km^2) land areas has received some form of protective designation. (For comparison, the cultivated area is 17.5×10^6 km^2.) However, this designation may not coincide with the inventory criteria since for example designated areas may well have roads in them since not all conform to the US definition. Comparing designated protected areas with the 1989 inventory is therefore only a very general guide, though it shows beyond any doubt that there is plenty of scope for more land to be scheduled for protection and to receive appropriate management. In Africa, for example, less than 7% of actual wilderness has protected status. It has to be realised, however, that designation does not necessarily mean that the area automatically becomes a true wilderness, especially away from high latitudes. As with all protected areas, a wilderness may exist in a matrix of more highly manipulated ecosystems and be subject to influences coming from them. A simple example is that of a migratory wildlife population which spends part of its annual round in the wilderness but another part in heavily used grassland where a wild herbivore is seen to be in competition with domesticated species for the available forage.

In reality, managers of wildernesses often have to accept that pre-existing resource uses may have to continue, either temporarily or on a long-term basis. Mining is one such activity: if it is the major conventional economic activity, then it may be argued that only the creation of wealth that it implies will bring enough income to manage the wilderness properly since that will inevitably be a charge on the public purse. In developing nations, the political influence of a large mining enterprise will likely be such as to sway governments in favour of them continuing their extraction programmes, no matter what a more conservation-minded regime would wish to do. The same is usually true of military considerations, where national 'security' normally has as high a priority in the land use sphere as it does in overall expenditure.

It is sometimes argued that the harvesting of renewable resources can be made compatible with wilderness designation. This is of course a paradox but such compromises do exist. Zoning, for example, as in biosphere reserves may allow some logging in fringe areas provided that proper re-afforestation is undertaken. Fishing in rivers, lakes or offshore in coastal wildernesses is an interesting issue. Strictly, this could be seen as similar to forestry but since the fish are not seen (i.e. are not part of most peoples' perception of the landscape, which is dominated by their visual sense) there is less pressure to end the practice, especially if it is part of an indigenous economy. More common in the developed world (though increasingly being extended to developing nations as well) is low-impact recreation in which walkers use the wilderness area but only in the company of what they can carry, or possibly from boats or rafts. 'Take nothing but photographs, leave nothing but your footprints' is the slogan (Plate 2.26); there are however some physiological difficulties in adhering completely to this.

Hence, active management of wildernesses may be necessary in spite of all that is implied by their designation. It may be necessary to manage animal populations, for example, if they are not completely self-contained within the boundaries of the designated area. Areas which were formerly hunted over may need the re-introduction of certain key species, like mammal herbivores or their predators. If the area is used for recreation then it may be necessary to have a fire control policy, even though natural fires are a normal feature of the ecology. Above all, it is usually necessary to declare that the area has a carrying capacity for people (even back-packers) which may not be exceeded. In fact, most wilderness management is recreational use management. Thus ecological and perceptual parameters have to be agreed: how much wear on footpaths is acceptable, for example, or which areas must be avoided entirely during an animal's breeding season? If a walking party sees another such group during the course of a week, is the wilderness experience destroyed? Local plans are therefore essential. The answers to questions of ecology may be scientifically determinable, but those of perception are culture-related: compare the tolerances to the nearness of other people in the outdoors of for example the Finns and the Japanese. In the end, a rationing system is often needed and the wilderness becomes a commodity to be booked rather like a seat at a concert.

No matter how tight the legislative Act and effective its enforcement, many areas cannot be isolated from all human-induced processes. A classic example is the tundra areas which form so much of the *de facto* wilderness of Fig 2.26. The fall-out from above-ground testing of nuclear weapons (a practice abandoned by the USSR and USA only after 1962

Plate 2.26 'Take nothing but photographs, leave nothing but your footprints' is easier in wild water rafting as in North Carolina. This type of 'wilderness' recreation has to be carefully controlled and 'rationed' by time, if the perceptual requirements of the participants are to be met.

and continued thereafter by China and France for some years) does not come to earth evenly. Atmospheric circulation patterns have produced a medium-term fall-out of radionuclides that has been heaviest over the high latitudes of the northern hemisphere. Further, since lichens grow so slowly, they have accumulated large concentrations of isotopes with long half-lives, such as strontium and caesium. A similar process has been characteristic of aerosol lead, which can accumulate in snow and ice (e.g. in the Greenland ice-sheet); though present elsewhere it is washed out into the oceans. And given the propensity of humans to leave detritus wherever they go (witness the pictures of climbers' litter atop remote peaks of the Himalaya), it is clear that nowhere, not even the wilderness areas, on earth is in fact beyond the reach of some trace of humanity.

The future for wilderness is always uncertain. On the one hand, an increasing awareness of environmental concerns and of the transforming impact of human societies provides pressures to set land aside for the untrammelled operation of natural processes. On the other, the current resource demands of developed countries and the expectations of the developing nations create an outreach for materials that can extend literally to the ends of the Earth: Antarctica, to be considered next, is an example of these tensions.

Antarctica

'Great God! this is an awful place', wrote the ill-fated Robert Scott in his diary shortly before his death in 1912. Yet perhaps because of the powerful environmental constraints exercised in this land of perpetual cold (Plate 2.27), it has come to play a role in resource

Plate 2.27 The wilderness aspect of Antarctica is brought out by this classic photograph of the *Terra Nova* in 1911, seen from an ice-cave.

issues little foreseen by the early explorers for whom it was merely a physical challenge to be conquered. Antarctica covers 14×10^6 km², which is about the size of Europe and the USA combined and has a substrate of rock under much of the ice (unlike the North Pole) and a series of sub-Antarctic islands. In winter, the ice-covered area extends to a total of 200×10^6 km². The continental territory is divided segmentally among 7 nations, but since the Antarctic Treaty of 1959 its management has been in the hands of the Treaty nations, which now number 16, with the original 12 retaining most of the decision-making powers. Although the land and ice surfaces are low in autochthonously nourished biomass, the seas harbour a diverse fauna and flora of plankton, crustacea, fish, whales and other marine mammals, and many species of birds, especially penguins. Some of Antarctica's resource and environment features are summarised in Fig 2.27.

The resource history focuses partly upon whales but also upon other mammals. In 1777 Captain Cook considered that 'the world will derive no benefit' [from Antarctica] but between 1778 and 1830 the demand for furs in Europe and North America resulted in the near-extinction of the Southern Fur Seal of South Georgia. Although these islands and the seas continued to yield marine resources, the Antarctic Treaty effectively froze all the prospects of resource exploitation in the name of science (in fact civil science, since all military activity was prohibited.) Since the treaty was catalysed by the international cooperation of the International Geophysical Year of 1957, scientific work has been the chief use of this continent since then, with work on geology and geophysics (including the discovery of the 'ozone hole'), the biology of the seas and the very harsh terrestrial habitats, and human biology in extreme conditions of climate and social isolation.

The Treaty did not however cover the offshore waters and so fishing and whaling has been pursued there as everywhere else, and only other measures have served to protect these stocks when their future has been endangered. Commercial cod fishing, for instance, was banned in 1985, under the provisions of the 1982 Convention on the Conservation of Antarctic Marine Living Resources (CCAMLR). At one time the krill (a small shrimp which features in many food chains) was being taken at the rate of 0.5×10^6 t/yr and so a limiting convention based on the whole ecosystem rather than the apparent MSY (see p 112–3) of the krill was effected in 1982 and the catch is now less than 0.2×10^6 t/yr. Another activity which post-dates the Treaty is tourism which is non-destructive if well controlled but nevertheless produces some environmental impacts.

Although the Treaty runs for an indefinite period, it is open for revision in 1991 and so concern has been expressed that in this demanding age, the pristinity of Antarctica will not be respected and the continent opened up to conventional resource exploitation. The minerals that are in view include the possibility of offshore hydrocarbons together with rock minerals such as chromium, copper, nickel and platinum which are found in the Dufek Massif. Although it is always difficult to make accurate forecasts, it seems likely that the costs of extraction in both cases would make enhanced recovery from more accessible places a better economic option. Greater threats to the integrity of Antarctic ecosystems probably come from the exploitation of whales should the IWC curbs end, uncontrolled taking of fin fish, and unbridled tourism. Offshore oil in an exploitable site might however be a different proposition. The Ross and Weddell Seas are estimated to contain recoverable reserves of $15,000 \times 10^6$ barrels of oil, comparable to the $20,000 \times 10^6$ of the British sector of the North Sea field and twice that of Alaska. There is a good deal of exploration of the legal kind at the moment about the basis of a régime for extracting all kinds of minerals from Antarctica.

Part of this threat has always been from accidental environmental impact rather than resource exploitation. The many vessels that supply the bases and that fish in Antarctic waters, for example, are a potential source of oil spills which would be very slow to disperse

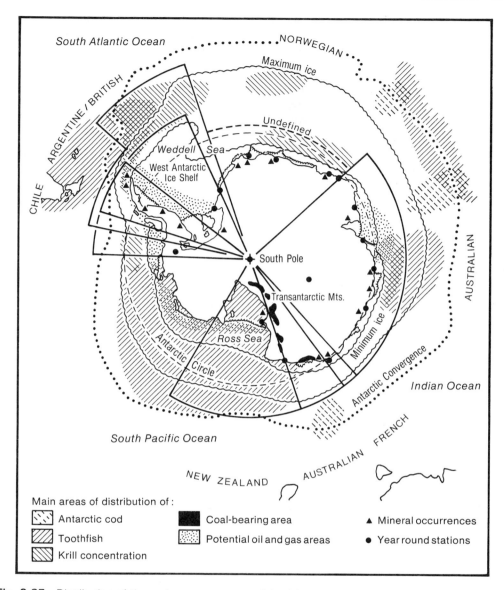

Fig. 2.27 Distribution of the major resource potentials of Antarctica, as evaluated in the mid-1980s. H. Zumberge, 'Mineral resources and geopolitics in Antarctica' *American Scientist* 67 (1979) 179–189; and N. Myers (ed) *The Gaia atlas of planet management*. London and Sydney: Pan Books, 1985.

and break down in these cold waters. Scientific bases are generally lax in disposing of the many solid wastes that they generate and some of them have disrupted the breeding grounds of birds. In recent years, for example, Greenpeace has accused Argentina and Uruguay of violating Treaty regulations by leaving around such items as loose rubbish like rubber, plastics and batteries. Since 1982, the French have been building (at Lion Island) an 1100 metre airship which has destroyed the nesting sites of at least 135 petrels and 3200 Adelie penguins.

Tourism has to be strictly controlled if penguin rookeries (to quote a single example) are not to suffer too much disturbance and cruise ships must not dump sewage or other wastes in Antarctic waters. The opening of the Antarctic treaty for amendment in 1991 has caused something of an international battle. The original signatories, which were the 'land-owners', later joined by India and Brazil, have been accused of forming an exclusive club which is a *de facto* last stand of colonialism and a political anachronism. Other nations (led in this case by Malaysia) wish to see the UN control the Antarctic as a 'common heritage for mankind', with a greater share in any resources that might be developed. As a complication, organisations such as Greenpeace see the stressing of the conservation issues as much more important. Whatever the outcome of these contentions, the Antarctic has played an important symbolic role in the ability of many often-opposed nations to cooperate in managing a territory even if, as cynics say, it was mostly ice, snow and birds that cannot even fly.

The role of the wild

This chapter has not all been concerned with the world's remaining wild lands but they have been a recurrent theme within it. To end it, though, it seems appropriate to consider some of the virtues of the little-manipulated and remote lands of the world. There is initially the use of the wild as a resource, which comes under two heads. The first of these is ecological in the sense that it contains species and ecosystems which are of economic value. Given that not all species are catalogued by science and that the characteristics of the named taxa are unevenly known, then the wild species of the planet must ineluctably be described as a treasure-house for humans. As well, many species and ecosystems are the very heart of the tourist attractions upon which many regional, and some national, economies depend.

The second we can describe as psychological, in the sense that there is a large body of argument which sets out the value to human societies of the presence of wild terrain; wild rather than necessarily natural is a useful distinction here. At one extreme, spiritual benefits are conferred by contact with nature and at the other, it is simply being somewhere different and away from the everyday pressures of life. That this is predominantly a feature of the industrial world is no surprise but there is sufficient evidence to suggest that many other individuals gain from a close aquaintance with the natural world.

But we must also consider the total role of the wild *in* the world. That this type of land is diminishing in area is probably as familiar to many of us from our own lives as it is chronicled by land-use statistics, which show the shrinking of areas of forest, natural grassland and wetlands, for instance. The inevitability of this process is also a normal part of our own mind-set, which is disturbed only when some undesired development threatens the habitat of the middle classes. Yet it is possible that wild areas play an important part in the regulation of certain planetary processes. The balance of the carbon cycle for example depends upon the sequestration of CO_2 by living organisms and this is fastest achieved by the natural vegetation on the land and the phytoplankton in the sea. So the maintenance of TMFs and shallow warm seas has a planetary role. Organisms living in the top few millimetres of the oceans give off dimthethyl sulphide (DMS) in aerosol form and these droplets from condensation nuclei which regulate the rainfall over the ocean and the down-wind continental areas. Thus living creatures unaffected by human actions (unless they are destroyed or otherwise, sub-lethally, affected), act as governors of key flows of a biogeo-chemical nature.

The ways in which conversion from wild to tame happen are many and various but they are fundamentally driven by resource consumption which in turn is related to population numbers. Not only the numbers by the level of consumption as well, of course, and in Part III we shall have to consider whether human population growth is indeed the biggest disrupting force in planetary function, as some have claimed it to be.

8

Wastes

The back-end of the resource process

Every living organism produces wastes which must be excreted in order that the individual's metabolism is not impaired. In natural ecosystems these materials are usually the food for another organism. The organisation of human societies has some analogies: in the course of using resources they produce wastes which are perceived as harmful to the body of society and which must therefore be disposed of. In many cases 'the environment' (i.e. the atmosphere, the seas, rivers and underground) is a place where the adage 'out of sight, out of mind' applies. In some cases these wastes are genuinely toxic to humans and other living organisms (high concentrations of sulphur compounds in gases for example), in others they are valued as undesirable by a particular culture, like pairs of old shoes. Generally speaking, the richer the society, the more easily are materials perceived as wastes; poorer people are apt to regard such disjecta as sources of raw materials. Wastes like any other material, therefore, are subject to ecological, economic and cultural appraisal by humans.

Given the high profile of wastes and waste management today, we are apt to think that it is only very recently that human societies have produced refuse. Any archaeologist will confirm that it is a long-standing habit (and indeed they would be in a bad way without it) but like resource consumption it has grown in scale as human numbers have increased and as their capacity to use more materials and energy has grown. At low concentrations, wastes can be ignored by people because they are not offensive or offer no threats to health or livelihood; in such circumstances we may talk of environmental *contamination* by humans since the flows of the natural world have substances added which would not otherwise be there or would be present in lower concentrations. At higher concentrations, wastes may cause damage to human health, to ecological systems, to the built environment and be offensive culturally. This level is generally labelled *pollution*. In addition, science and technology can produce substances not present in the natural world and whose side-effects in the environment are not always predictable. Its manifestations today are complex so that unforeseeable interactions between contaminants can occur in the process called *synergism* (Plate 2.28).

Plate 2.28 Mexico City. In this basin, photochemical smog builds up and obscures visibility. Less obvious from a photograph like this is the damage done to plant, animals and materials.

Classifications of wastes

We can bring to this task a number of different points of view (Table 2.28). One would be the everyday perception by individuals or a group of what is a reasonable risk from a toxic substance or a tolerable level of contamination of their immediate environment. Thus people living near a major chemical works, for instance, have decided that the risk of a major leakage of a toxic substance is low and also that they do not mind the noise of the traffic to and from the plant. (This assumes that the inhabitants have a choice, which is unlikely if they are poor.) Noise is in fact a good example of a contaminant whose effect is largely determined by cultural and individual variables. Above certain levels it can produce deafness but below that individual tolerances are very wide and indeed it is possible to become an addict at certain levels: travel in any train with a leaky-walkman owner.

A second classification is commercial, operated alike by companies and nations. This evaluates wastes as sources of profit for industry. There are two main areas of potential money-making here: the most obvious is in the reclamation of useful materials from wastes as in the scrap metal industry or in the intermittently gainful waste paper business. The second is in the transport, treatment and 'ultimate' disposal of toxic materials on behalf of other industries or society at large. Thus toxic substances produced by an industry as by-products or which are exhausted components of an industrial process and which have no possible further use may be sold to a reprocessing company. Toxics like PCBs and dioxin are then taken to countries which have advanced destruction facilities such as

Table 2.28   Ways of classifying environmental contaminants

(A)	By nature
	Chemical composition e.g. inorganic, organic
	Physical state e.g. gaseous liquid (pure, diluted), solid (pure, suspended)
(B)	By properties
	Solubility in water, oil, fat
	Rates of dilution and dispersion
	Biodegradability
	Persistence
	Reactivity with other substances (synergy)
(C)	By sectors of environment
	freshwater, marine, land, air
(D)	By source
	Fuel combustion
	Industry
	Domestic and institutional
	Agriculture
	Military
	Fermentation processes
(E)	By patterns of use
	Industry
	Domestic and institutional
	Agricultural
	Transport
	Defence
(F)	By effects on 'target'
	Atmospheric process (e.g. ozone, CO_2 concentration)
	Corrosive agents
	Direct effects on livestock and wild animals
	Direct effects on crop and wild plants
	Direct effects on ecosystems

Source: M. W. Holdgate, *A perspective of environmental pollution*. CUP 1979.

high-temperature incinerators and reduced to simpler and less poisonous compounds. This may be done at sea to lessen the risk to human inhabitants. Alternatively, such substances may be off-loaded onto a developing nation which is not so sophisticated in its requirements for treatment and where profit margins may thus be higher. Commercial attitudes extend all the way up to the reprocessing of nuclear wastes containing plutonium, possibly the most toxic substance in any resource process. Plans to fly such wastes from Japan to the UK, for instance, are put forward from time to time, and the railways of nuclear power-possessing nations regularly transport these materials. In some countries reprocessing is in the public domain, paid for by taxes but in most it is also regarded as a source of potential profit.

The third way of classifying contaminants and pollutants is scientific, based on the objective measurement of quantities, concentrations, locations and flows. This is applicable to all types of wastes and is the best basis for economic and cultural evaluations, although complete data are rarely available because of the cost of monitoring, which is nearly always borne by the public purse. It is, though, an absolute necessity at all scales since it alerts us not only to local problems but to potential changes on scales which can affect the whole planet.

Scientific classifications

Table 2.28 gives us a set of choices about types of classifications of wastes and immediately creates a problem in the sense that they all overlap and there is no space to describe each one separately. Here we shall concentrate on a brief description of the kinds of wastes produced, followed by a longer examination of the main environments into which wastes are emitted and the interactions therein. Ideally, the first possibility demands a catalogue of the nature and amounts of all the various types of substances and their states which become wastes, at the point at which they become identified as wastes rather than useful materials in a production or consumption process. If this is done, then there is the virtue of comprehensiveness, but little is said about the fate of the various materials. Nevertheless a preliminary sorting into solids, liquids, gases, and radioactivity, together with chemical composition, is always useful. Intermediate states can exist (like aerosols) and some substances can appear in different states (sulphur for instance can appear as a gaseous oxide, a liquid acid or as a solid). Equally critical is the lifetime of the waste substance: does it undergo breakdown to different compounds which are less or more toxic, can it accumulate in food webs, might it be transformed in a synergistic reaction? Most industrial firms in the developed world are required to know what substances and in what quantities are being emitted from their plants. Only in the most advanced is this public knowledge since commercial confidentiality is often used as a cloak for keeping the public less than well informed.

More common is the practice of classifying the process of waste disposal in terms of the sub-system of the environment into which it is led. This may lead to repetition in summary accounts but has the great virtue that it deals with the fate of the substances which form the emissions. Naturally enough, it may be very difficult to monitor accurately all the substances everywhere all the time (where is all the DDT and its metabolites that have been flushed into the seas?) but even trace quantities in the upper atmosphere or in the deep oceans may be significant and ought not to be overlooked in favour of the more obvious. The major compartments into which wastes are led (land, air, rivers and oceans) are dominated by three which are usually public property (unlike the land) and so the results of monitoring are often known, except that they may again be kept secret for commercial or political reasons. Lastly it is clear that these compartments are not 'watertight' and that transmission of contaminants between them takes place so that the pathway of a particular substance may need an integrated system of monitoring if its true fate is to be known and evaluated for risk to ecosystems or directly to humans.

Major waste flows in the environment

Any scheme which treats wastes by compartment is bound to be imperfect because of the way in which they transgress sub-systems. Lead from antiknock additives in petrol, for example, contaminates the air, the land around roads, water, and falls out from the upper atmosphere onto high-latitude ice sheets. It is however most significant when it accumulates in food chains (vegetables from a garden or allotment near a busy road are at risk) or is directly ingested from paint by children. So a complete account of all pathways for all substances is simply not possible.

Pollution of the land

Since this is where humans live, it is no surprise that concentrations of contaminants are high here, though strenuous attempts are made in most societies in the West to lead them off to less 'visible' environments. Some emissions are, of course, not detectable by the unaided senses: radioactivity is probably the most important of these. Above-ground testing of nuclear weapons has left a legacy of on-land radioactive isotopes in high-latitude tundras, where bioaccumulation in lichens and herbivores has occurred. Accidents like the Chernobyl partial meltdown of 1986 have also contributed to pollution thousands of kilometres away from the actual site as well as sterilising the immediate surroundings for many purposes. (Like habitation: 235,000 people were eventually evacuated.) A recent addition to the repertoire of invisibles has been the possibility of deleterious effects on human health of the electromagnetic emissions from high-voltage powerlines to those living close by. As with many such non-obvious effects, definite relationships are hard to prove. No such difficulty exists with noise and vibration, however, where precise scales of damage to hearing from elevated noise levels are known, and the overall effects of vibration (e.g. from heavy vehicles) on structures and people are closely documented.

Mention of radioactivity reminds us that nuclear power stations have a limited life (of the order of 25–40 years) and that the structures are heavily contaminated with radioactivity. They must be disposed of either by being dismantled and the parts treated as other nuclear wastes (see pp 65–6) or by being entombed in concrete on-site. Eventually the sites will be safe to re-use, but the isolation time will be of the order of hundreds of years: it seems unlikely that the mounds will be available for use as artificial ski-slopes. Monitoring will be absolutely vital in these places, which will be quite plentiful elements of the land use pattern of even partly nuclear-powered nations. The actual reactor is not the only part of the cycle which has to be decommissioned and hence decontaminated. Fuel manufacturing plants, enrichment plants, storage ponds and reprocessing plants will all need such treatment at the end of their working lifespans. Even the decontamination centres of the plants have to be decontaminated and dismantled. In total, decommissioning of a reactor is estimated to cost between 20–40% of the capital cost of the reactor. For the UK's 26 Magnox reactors, a current estimate is £6000 million.

Many solid wastes are tipped onto land that can be spared from other purposes. This is especially true of mines and quarries, but is also found at other industrial sites and is especially so of household and institutional ('municipal') solid wastes, whose composition reflects in both quantity and quality the contributing society. As an instance, New York City generates 30,000 t/day of domestic waste which would include such items as the 30 large plastic sacks needed to contain the waste generated during each routine thoracic operation. In fact the London hospitals generate enough clinical waste each year to fill Trafalgar Square to the height of Nelson's Column three times over, i.e. 33,000 tonnes/ year. Perhaps the most obvious resultant pollutant is dust, which can be a hazard to health and safety as well as diminishing amenity: asked anybody living near a steel plant or an open-cast mine. Garbage tips, if properly managed, may do little harm and eventually provide a land surface suitable for e.g. playing fields or afforestation though not for load-bearing structures nor housing. But they must be water-tight at the base and each day's debris must be filled over with soil. Failure of the first condition means that toxic wastes can leak into surface or ground water; of the second that a habitat is provided for rats and fermentation bacteria that will generate methane that can catch fire. Heavy rain can cause site effluents to overflow, and most liners of clay or plastic will eventually leak. The Lipari toxic waste site in New Jersey seems also to be the centre of a leukemia cluster. Controlled production of methane is a useful source of energy but this requires even more careful management of the rubbish tip. Overall, disposal sites for solid wastes are undesirable neighbours, though their resources are often sought after by the few who can make a living from unofficial or sanctioned recycling and re-use of discarded materials.

A last general category is that of derelict land produced by industries, often now vanished. It may simply be morphologically difficult to use, e.g. full of holes and heaps, or may be toxic: with high concentrations of metals like lead or copper. Reclamation is usually possible and such land can be restored to productive use of many kinds, though the costs of doing so to society are high if the exploiters of the original resource have not been made to pay for restoration. Most advanced countries now impose restoration conditions and costs on developers, though some poorer governments are unable or unwilling to do so.

Productive land may contain all kinds of residuals which, although not immediately toxic or offensive, nevertheless form a pool from which they can be taken up into food chains. These chains may lead to humans and concern has been expressed from time to time at the levels of some persistent chemicals in human diets, though more anxiety is currently expressed over additives from the processing and storage stages. Nevertheless, the Polish government has had to consider a ban on vegetable growing in several Silesian towns where the soil concentrations of cadmium, lead, mercury and zinc are 30–70% higher than the WHO norms.

More typically, substances such as organochlorine pesticides (of the DDT family for instance) break down very slowly in soils and very little at all in animal metabolism or storage in fats. When they do break down, their metabolites may in some cases be more toxic to life than the original substance. Heptachlor becomes heptachlor epoxide and then heptachlor epoxide ketone, each of which conversion increases toxicity. The result has been a series of lethal and sub-lethal effects upon wildlife in agricultural areas; this is paralleled by rivers and the sea where these substances have been transported by run-off. Animals vary in their ability to excrete organochlorines and the substances may thus build up in their fatty tissues. Levels may become lethal (sometimes this happens when fats are broken down and release the toxins into the bloodstream and this is likely to be when the animal is under-nourished and so resistance to stress is low) and kill a proportion of the population, though there are usually some resistant individuals. Sub-lethally they may depress the level of reproductive hormones or in the case of some birds induce thinning of the eggshell so that few eggs hatch successfully. This phenomenon has been noted especially in predators such as sparrowhawks and peregrine falcons, whose populations have recovered if this class of pesticides has been replaced with less persistent chemicals. (We might remember, too, that declines in songbird populations from this cause were the trigger for Rachel Carson's book *Silent Spring* (1963), which was catalytic in raising environmental consciousness among western societies.)

Outside the close control exerted on pesticides in the West, there is a circle of poisoning. DC firms still export these substances to LDCs, where both workers and environments are contaminated. Imported cash crops then exported to DCs are then high in pesticides banned in the importing country; testing is not very thorough: one sample of coffee beans in the USA contained six different pesticides banned in the importing nation.

Industry may also contribute similar materials: lead has already been quoted. Once an additive to paints, and thus a hazard in areas of older housing as it breaks down or, apparently, is chewed by children, it is now most common (a) in old water pipes and (b) in aerosol form from motor exhausts. Happily, both sources are diminishing in quantity as pipes are replaced and lead-free petrol becomes more acceptable to consumers. But lead concentrations are high in urban areas and near roads, especially where it can build up in slow-growing parts of plants like fruits. Here then is a direct pathway to human ingestion, and above certain levels lead can act as a poison of the central nervous system, though rarely lethal. But it is no comfort to know that in the DCs body burdens of lead today are 500 times those of most pre-industrial populations.

We would all like the land where we live to be only a temporary resting-place for wastes but enough has been said to show that, alas, its biochemical nature combined with human habits ensures that a great many substances stay around long enough to cause changes in the metabolism of individual animals (including our own species) some of which are bound to have negative effects on well-being.

Freshwater

Rivers are an especially good place to dump wastes, since the flow of water usually carries them away to be diluted or to be somebody else's problem; land burial involves more effort and tides have the bad habit of bringing the stuff back again. So it is no surprise to find rivers and lakes are the recipients of many wastes and often so changed that their original biota are no longer present. The Hangpujian River near Shanghai, for example, has one volume of untreated waste for every 4–6 volumes of water, and 54 out of the 78 major rivers of China are said to be seriously polluted.

One of the most pervasive of water contamination processes is that of human-caused nutrient enrichment, or *eutrophication*. Rivers vary in their natural concentrations of elements like nitrogen and phosphorus but low levels are common since the flora sequesters them. Hence, runoff from agricultural land or from sewage which contains high N_2 and P levels is likely to lead to biological changes. The most common, as in the sea, are 'blooms' of algae which can reproduce by simple cell division but whose populations are normally limited by the low levels of nutrients like N_2 and P. Released by the contaminating supplies, the algae grow rapidly but as they die, bacteria also bloom and take up much of the oxygen in the water. If the water is warm the oxygen levels are depressed anyway, and so fish die quickly and sometimes in large numbers. In the USA, for example, farm runoff affects 64% of all rivers and 57% of lakes. In Europe, the levels of N_2 have been increasing in many rivers; the WHO recommended limit is 11.3 mg/l but one-third of all UK rivers exceeded this in 1978–88.

Sewage may be subject to varying levels of treatment before it is released into rivers or lakes. This rarely takes out the N_2 and P, but the use of chlorine usually kills most of the bacteria which otherwise will cause disease if the water is used for drinking and cooking without being boiled or otherwise disinfected. The index of contamination is the level of coliform bacteria present in water and in India, for example, the Yanuna River before coming into New Delhi has had 7500 coliforms/100 ml water, and below the city, 24 million per 100 ml water.

Sewage disposal is an enormous problem in large, growing and poor cities. Cairo in the 1970s, for example, was regularly flooded with untreated sewage and today it treats less than half of the 2 million m^{-3} that enters its disposal system. An immense project to improve the collection and treatment (and to produce water and fertilisers) is currently under way.

If the runoff is contaminated by elements or compounds in solution then it is likely that groundwater will reflect such flows. Soil or rocks may exert a filtering effect though in the UK levels have risen by three times in the last 20 years. For N_2, the US recommended limit in wells is 10 mg/l but an estimated 8200 wells exceed this limit; the effects of high levels of N_2 on adults are unknown but babies are affected by 'blue baby' syndrome at 745 mg/l. Pesticides are also found now in many aquifers: aldicarb for instance in 40 states in the USA: in LDCs many more would probably be detected. Groundwater also receives leached nutrients and compounds from cesspits and leaky landfills: as many as 7300 landfills in western Europe are estimated to need immediate action to contain contaminants of various kinds.

Ever since about 1800, power plants have been increasing the amount of sulphur emitted into the air, and much of this has fallen out in acidic form. It can easily be exported: the UK exports 3.7×10^6 t/yr and the USA, 24×10^6 t/yr. While Sweden's deposited load is 90% from abroad, in turn it exports two-thirds of its own emissions. In lakes and rivers, the effect is to lower the pH, with consequent impacts upon the flora and fauna of all kinds (some lakes showed changes in diatom floras as early as 1850) but especially upon commercial and sporting fish populations. The exact course of the acification is not predictable since the ecology of the entire watershed exerts influences upon runoff (the soil acidity level is an obvious variable) but many freshwaters in industrial nations formerly at pH >6.0 are now in the 4.8–5.5 bracket. The changes are especially noticeable in Scandinavia, the north-east USA and south-east Canada, and Scotland. In southern Norway, for example, 1700 lakes have lost their fish populations.

The consequences of acidification of these waters are far-reaching: for instance the decomposition rates of lignin and cellulose are reduced because of lower levels of bacterial activity, so plant debris begins to accumulate on lake bottoms and become decomposed only by anerobic bacteria so that methane and H_2S are given off. The biota of river substrates are especially vulnerable and so snails, clams and crayfish disappear. The lower pH accelerates the rates of leaching of some metals into rivers whence they may be found in drinking water: pH levels of 4.5 have been found in some wells, along with detectable concentrations of lead, mercury, cadmium, aluminium and cobalt. Aluminium especially has been implicated in the death of fish. This is especially so in northern and upland climes where a spring flush of meltwater is especially high in acids accumulated in the snow during the winter. The pH can fall rapidly in a few days and larvae and fry may be unable to survive the stress. It is also thought to be causative in Alzheimer's disease in humans. And lastly, high acid levels can speed up the corrosion of structures in water just as on land.

Reversal of these trends has to begin at source and it is possible to scavenge out much of the sulphur in power station wastes, though not without cost. In Europe, most nations have committed themselves to a 30% reduction in sulphur emissions by the end of the next decade, with an increase of about 6% in the price of electricity. Central Europe has the greatest task since lignite is such an important energy source and it is very rich in sulphur; natural gas on the other hand is the reverse case. On-site, lake acidification can be reversed by liming, though this is expensive and thus available only in the richer nations. Sweden, for example, has planned to lime 3000 lakes in the first 10 years of a programme during the 1970s and 1980s.

Acid precipitation (Fig 2.29) (often popularly referred to as acid rain) is a very high-profile contaminant which affects the land was well as freshwaters. Since buildings and forests are so visibly affected (though their corrosion and deaths are probably the outcome of the interaction of sulphur compounds with other emissions as well), public interest has been easily aroused and so a positive feedback is well under way in terms of sulphur output reduction. Nevertheless, acid precipitation is not a global issue: it is mainly confined to those areas that (a) can detect and measure it accurately and (b) can afford to clear it up. In this respect it is instructive to compare it with other contaminants which are more difficult to keep track of and so expensive to deal with that for example no developing country could begin to think of combatting any problems that arose.

Water is a convenient medium for the disposal of various toxic or noisome wastes Plate 2.30. Solids, liquids, tailings, silt, chemicals, oil and agrochemicals can all find their way into watercourses by accident or design. Accidents have included the very chemicals

Plate 2.30   In 1986, a government-owned brewery in Taiwan released a toxic substance into this river, killing about 40 000 kg of fish. In the hot humid climate, these decay quickly, adding to the levels of toxicity and contamination.

that in small doses are used to purify water at treatment plants. In many industrialised nations, regulatory mechanisms based on legislation are used to try and reduce such flows of wastes and indeed the lead content of the Rhine, to quota one example, fell from 24 μg/l in 1975 to 8 μg/l in 1983. The mechanism of achieving such falls is variable from nation to nation and often has an ideological content. The stronger the socialist bias in government then the more likely it is that regulations will be used; at the other end of the spectrum, free-marketeers favour systems of licences to pollute which can be bought but whose cost makes it cheaper for a firm or individual not to pollute in the first place. Both these processes depend on the presence of an independent monitoring organisation paid out of public funds which can provide scientific data, and on pressures from the general public which in turn needs access to objective data. Such data are often kept secret however: many Stalinist administrations fear the political consequences for themselves of the truth and many capitalists are in thrall to the idea of commercial secrecy.

Probably one of the oldest contaminations of river water is that of sediment, which must have been changing water quality ever since agriculture and pastoralism began. Yields to runoff are especially heavy in areas of intensive agriculture, of deforestation (notably where slopes are steep), and of rapid urbanisation or industrial construction. In a country

Plate 2.29 Acid precipitation in its various forms is known to produce die-back in trees. This is demonstrated in a most heavily polluted area of Czechoslovakia in the 1980s, but similar scenes could be found in much of eastern Europe and Germany at that time.

like the USA, sediment is the primary pollutant of 47% of the rivers, measured as a proportion of the length of rivers classified as polluted in some manner. Heavy sediment loads change the ecology of rivers very radically: the increased total volume increases the likelihood of flooding; the finer material blankets and kills much animal life (especially fish), and suspended matter inhibits photosynthesis. Ubiquitous though sedimentation is, it is one of the most difficult to control since it derives its effect from the whole pattern of land use.

Like the sea, rivers are also subject to manipulation from the input of large volumes of hot water, called *calefaction*. More so, in fact, since the ratio of heated water to receiving fluid is greater. Hence, water plant and animal species can be much affected and there are many instances of escaped tropical fish populations becoming established in the calefacted zones of temperate rivers. Stories are told, but rarely confirmed, of giant alligators inhabiting the rivers of Germany.

An overall view is difficult to establish and the situation is changing rapidly in many nations, for better in some and worse in others. A gross generalisation might be that point sources of contamination are being subject to clean-up and improvement but the non-point sources (especially nitrogen and sediment) are not, and at natural gas is the reverse case. On-site, lake acidification can be reversed by liming, though this is expensive and thus available only in the richer nations. Sweden, for example, planned to lime 3000 lakes in the first 10 years of a programme during the 1970s and 1980s.

The oceans

There are several main kinds of wastes which reach pollutant levels in the seas (Table 2.29). For most of us, the obvious ones in terms of personal experience and media attention are sewage and oil. Sewage is the main example of an oxygen-depleting waste to affect the sea, but others in this class include other organic materials (like the effluent from pulp and paper plants), and eutrophicants like phosphates and nitrogen. Sewage is either fed raw into the sea by means of offshore pipelines of varying lengths or dumped as sewage sludge, which is the dewatered solid (and hence very concentrated) residue after treatment. Since 1000 people will give rise to 25 t/yr of solids, large areas of offshore waters can be affected. The major effect of these substances is to provide a substrate for bacteria which in the course of their growth take up some or all of the dissolved oxygen in the seawater. If the oxygen is all consumed then anaerobic bacteria may flourish, releasing gases such as methane, hydrogen sulphide and ammonia. Oxygen-free zones are usually free of fish since these animals can escape to other areas, but retain phytoplankton since there are good supplies of nutrients and carbon dioxide. Any remaining fish get their gills choked by plankton and there is a heavy rain of the plankton onto the benthos where they choke most of the living organisms there. Good supplies of nutrients also make for 'red tides' which are blooms of dinoflagellates whose excretions are often toxic to filter-feeding organisms such as molluscs. Raw sewage is also a plentiful source of coliform bacteria and parasite eggs which can affect human health: we may be taken short by the first but the second are likely to last longer. The effect of raw sewage and accompanying solids upon the amenity value of the seaside is also unpleasant.

Oil gets into the sea anyway from natural seepages: perhaps 0.5×10^6/yr of it, to which is added 1.38×10^6/yr from the human economy, which is not surprising since 50% of the 3.0×10^9 t/yr of the oil consumed is transported by sea. The most conspicuous form of pollution is the major spill from a tanker accident but there is a continuous

Table 2.29   Major pollutants of the oceans

Pollutant	Toxicity to organisms	Hazards to human health	Reduction of amenity value
Domestic sewage	+ +	+ +	+ +
Organochlorines	+ +	?	
PCBs	+	?	
Acids and Alkalis	+		
Metals e.g. Mercury	+ +	+ +	+
Lead	+	+	
Radioactive isotopes		+	
Oil	+		+ +
Plastics	+		+
Pulp and paper wastes	+ +	?	+ +
Military wastes	+		+
Heat	+		
Solids	+		+ +
Dredging	+		+

Note: Considerable generalisations are made e.g. metals like mercury are much more toxic than zinc or copper

Symbols: + Significant   + + Important   ? Uncertain

Source: M. W. Holdgate, *A perspective of environmental pollution*. CUP 1979.

flow from minor spills at terminals, tank-cleansing, off-shore production and effluent from refineries. When a major accident occurs, various measures are deployed: floating booms, slick-lickers and dispersant chemicals are used at sea, and on the beach steam hoses and more dispersant can be used on hard surfaces whereas sand has simply to be scraped up by hand or machine. The Exxon Valdez spill in Prince William Sound in 1989 covered 25,000 km^2 of coastal and offshore waters and killed between 100,000 and 300,000 birds of 90 different specics (Plate 2.31)

Both the spill and the cleanup produce environmental effects. Some of these are usually short-lived: plankton, fish and tourist populations are not usually affected for many months but birds, especially those with a low reproductive capacity, may undergo declines lasting years. Shellfish beds are usually wiped out and recover only slowly, and fish in the area of a spill may be tainted with an oily flavour and hence unmarketable. If sub-littoral herbivores such as sea-urchins, limpets and abalone are killed (often as much by a dispersant as by oil) then large algae like sea-wracks smother everything until the herbivores recover their numbers.

Metals are natural constituents of sea-water, to which they make their way from volcanic eruptions, dust, the erosion of rocks, smoke and decayed vegetation. Very many human activities add to the mobilisation rate of metals, which in total increase the metal loads of rivers and the atmosphere, as well as occurring in direct discharges of, for example, sewage. The metals of most concern as contaminants are

- Transitional metals (like iron, copper, manganese and cobalt) which are essential for life in low concentrations but toxic at higher levels.

- Heavy metals or metalloids (e.g. mercury, lead, tin, selenium and arsenic) which are not used by living organisms and so are toxic to them at low concentrations.

Organisms vary in their capacity to regulate these substances in their bodies: plants and molluscs are generally poor regulators whereas crustaceans and fish can manage their levels of e.g. zinc and copper though not the heavy metals.

Mercury (Hg) provides a good example: natural inputs from rock weathering and geological degassing amount to *ca* 28 500–153 500 t/yr, and humans are responsible for adding about 8000 t/yr, some via the atmosphere and some as industrial effluents via runoff into estuaries and inlets. Microbial systems in the seas and estuaries can covert inorganic mercury into methyl mercury, which is (a) readily accumulated by living organisms and (b) much more toxic than the inorganic forms. Thus most sea creatures have some Hg in their tissues, though in areas of high input, levels are much higher. The mercury can also accumulate in food chains, so that in Minamata Bay in Japan in the 1950s plankton contained 5 ppm but fish 10–55 ppm. Human deaths (43) and chronic disabilities (700 permanent cases) resulted since mercury affects the brain functions: the syndrome is now known as Minamata disease. Standards have now been adopted in Japan as elsewhere which limit the amount of Hg in seafoods for sale: these are mostly between 0.5 and 1.5 μg of Hg per gram of fish or other seafood.

The class of chemicals known as halogenated hydrocarbons are, like metals, permanent additions to the biosphere. Unlike oil, however, they undergo no degradation from oxidation or bacterial action and are thus able to contribute to bioaccumulation if conditions are right. They are substances unknown in nature: totally humanly-produced, and can only be destroyed by incineration at temperatures above 1200°C. The two main groups are pesticides and PCBs. The pesticides include the well-known DDT group as well as aldrin and its relatives, lindane and toxaphene; PCBs (Polychlorinated BiPhenyls) are industrial

Plate 2.31   The *Exxon Valdez* in March 1989, surrounded by some of the large quantities of crude oil which escaped from her hull when she hit rocks off the coast of Alaska. Before the Gulf War of 1991, this incident was the world's largest oil spill.

chemicals renowned for their stability. Most of both sets are under careful regulation in the industrialised world but in the South they are still widely employed: DDT against malaria for example. Even so, large quantities of both are present in the environment, including the oceans. Disposing of used PCBs is often done by incineration at sea (due to be phased out in the North Sea by 1994) which produces carbon dioxide, hydrochloric acid and metals, though some traces of dioxin are said to have been detected.

As on land (see p 160–2), bioaccumulation can occur, with progressively higher concentrations of halogenated hydrocarbons up the levels of a food chain. The concentration factor between seawater and an organism is often of the order of 40,000–700,000 and so metabolic effects have been noted. In phytoplankton, for instance, NPP rates are reduced at quite low aqueous concentrations of DDT. In sea-birds, eggshell thinning is common in areas of DDT concentration in sea-water. In other marine populations, deaths and disease epidemics have occurred from time to time in areas known for their halogenated hydrocarbon levels but causal connections have never been established, in part because most marine populations seem to fluctuate with a wide amplitude under natural conditions.

Effects upon human health, likewise, are hard to document with certainty. PCBs caused some human poisoning at Yushima in Japan in 1968, producing all the symptoms of acne but other cases are debateable. Control of DDT levels in foodstuffs has meant that at times (especially in the late 1960s) human milk would not have been passed for sale in the supermarkets of North America. The effects upon today's 20 year-olds are not immediately apparent. Nor can we make confident predictions about the halogenated hydrocarbons still present in the world's oceanic ecosystems: dilution may render them harmless to life, or bioaccumulation may mean they are a kind of chemical time-bomb. Leakage of PCBs out of landfills in the UK (perhaps 600t/yr) has been implicated in a drastic reduction of other populations.

Radioactivity is also present under natural conditions. Average levels would produce sea-water with an activity of 12.6 Bq/litre, sands with 200–400 Bq/kg and muds with 700–1000 Bq/kg[1]. Radioactive substances do decay (the time-span of which is given by the measurement of the half-life) but they may still be very long-lived and different kinds of radiation do variable damage to living organisms. The additions made to these levels by human activity come from military testing in the atmosphere, now stopped, planned releases from nuclear power and reprocessing plants and the dumping of solid radioactive wastes. Some radionuclides remain in seawater (e.g. caesium-137) whereas others are sequestered onto sands and muds: plutonium-240 is an example of the latter tendency.

No known effect of radionuclides on wild populations of plants and animals has been proven under normal conditions of discharge from civil installations. That bioaccumulation occurs is acknowledged and emission limits are set which aim to keep body burden levels well below the international standards for various radionuclides. Potential pathways to humans are also closely scrutinised: heavy fish eaters round the Irish Sea are reckoned to be a critical pathway for the discharges from the Sellafield reprocessing plant, for example. The adsorption of plutonium onto sediments means that particles can be blown onshore from beaches at low tides, forming another pathway. In spite of a great deal of monitoring and of close study, the fear of childhood leukemia clusters round nuclear plants still persists and until these are shown to be epidemiologically unsound or a causal link is discovered, then emissions of radioactive waste to the sea (as elsewhere) will generate much public suspicion.

1. Bq = Bequerel, the unit of measurement of radioactive decay. One Bq = 1 disintegration per second. It replaces the Curie (1 Bq = 27–03 PCi).

Solid wastes have always been dumped in the sea, and strand-lines have exhibited evidence to the fact that they have not always been borne away. Today, for example, sea-bed dredging creates very high burdens of particulate matter which can blanket bottom fauna but release large quantities of nutrients at the same time. The effects therefore are complex but undeniable in the short-term; some idea of intensity may be gauged from the fact that 30–35×10^6 t/yr of sand and gravel are dredged from the North Sea alone. The east coast of North America has similar impacts. Mining wastes, too, add to the silt burden offshore: such spoil may be dumped directly into the sea, carried out by barge, borne out by rivers or even piped out to sea as a slurry. Its impact is immensely variable according to the vagaries of substrate, currents and tides, depth and season but is never negligible. Where metal ores have been sought then it is always likely that relatively high concentrations of metals will be present in the wastes and they will prove to be toxic to organisms. The newest member in the family of added solids is plastics, as a walk along most beaches in the world will now reveal. Most of the non-degradable bottles, for example, come from shipping, but industry is the source of the many small pellets (3–4 mm in diameter) of partially broken-down plastics which are found in most oceans both near and far from land. Another material which is largely 20th century in origin is munitions which have been dumped at sea because they are obsolescent or have sunk along with the ships carrying them and are too dangerous to move. They may not change the ecology directly but are of course a hindrance to fishing and sometimes even to navigation; in European waters they are surprisingly frequent (Fig 2.28).

Last of this catalogue of contaminants of the sea is the addition of heat, mostly in the form of a plume of hot water from power stations, probably 12–15°C above the ambient sea temperature. In cool waters it mixes with the receiving water and this results in rises of perhaps 0–1.7°C of the seawater. In warmer waters, living things may be nearer to the upper tolerance levels for temperature and so the impact on life is greater: around a North Sea or North Atlantic outfall, less than 1 ha may be affected biologically whereas in the subtropics up to 40 ha has been recorded. Life may also be affected by chlorine and leached metals in the outflow water.

One general observation may be made: the vulnerability of estuaries to change as a result of almost all of these contaminants. Because they are shallow, because they accumulate sediments, and because the tides may produce low net outfall rates of water (along with its burdens of contaminants), estuaries are prime sites for pollution. But under more natural conditions they also have a high biological productivity and this shows for example in their role as breeding grounds for many commercial fish populations. If we look, then, at the places in the world where the seas have undergone most change at human hands then the estuaries can claim that position.

Contamination of the atmosphere

The atmosphere is a complex system whose dynamics are only partially understood and for which accurate prediction is often impossible, even at the level of short-term weather forecasts. So the pathways and concentrations of contaminants are often difficult to trace and to predict, even more so than in the oceans. Nobody minds this very much as long as there are no deleterious effects upon humans or upon the ecosystems which provide us with resources: that is to say it is permissible to contaminate the atmosphere so long as there is no fall-out with an impact, direct or indirect, upon ourselves. As we know now, most contamination of the gaseous envelope of the Earth does in fact come back to affect us,

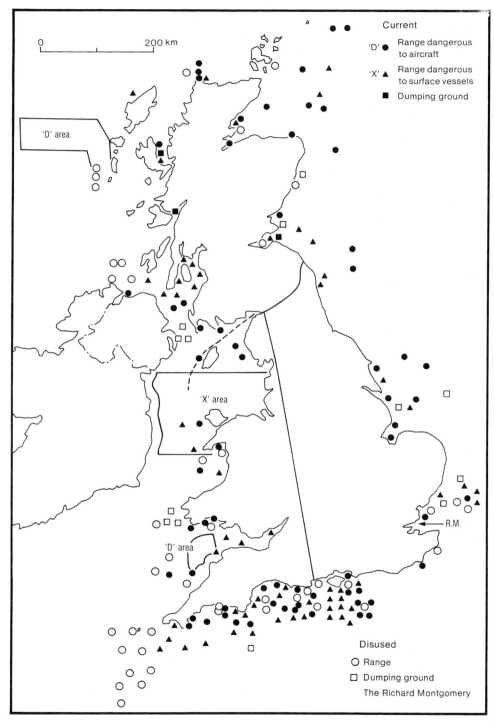

Fig. 2.28 Current and disused areas of UK coastal waters with a hazard from munitions. R. B. Clark, *Marine pollution*, Oxford: Clarendon Press, 2nd edn 1989.

in a number of ways. One measure suggests that in DCs air pollution causes damage to the tune of 1–2% of GDP.

Though pollutants are often found together or act synergistically, it will be convenient to classify them according to their nature at the points of emission: we will distinguish *particulates*, *gases* and *aerosols*, *heat* and *radioactivity*. There will be some emphasis on gases since these are probably the most important in terms of global rather than regional effects and also in implications for many aspects of planetary biophysical and cultural systems.

Particulates (more properly called Suspended Particulate Matter or SPM) consist mainly of very small pieces of carbon, hydrocarbons or sulphur compounds with a diameter between 0.1–25.0 μm. These are given off by a number of natural processes such as windflow over deserts and the oceans, volcanoes, forest fires, and soil erosion. Human eoconomies add to these in the form of fuel combustion, fires, ploughing and the creation of other unvegetated land, and industrial emissions. The natural processes add some 1320×10^6 t/yr to the atmosphere, and estimates of the human-led emissions vary from $60–300 \times 10^6$ t/yr, but at all events a much smaller source than those of nature. The majority of SPMs fall out at relatively short distances downwind from the point or area of emission and so with the human-induced sources there is a directly observeable link between deleterious effects and their immediate cause. Thus damage to all kinds of materials, to visibility and to human health can be directly traced to the kind of emissions which typically produce concentrations of 20–100 μg/m^3 in cities and under 10 in rural areas. Calcutta in the 1970s reached 360 μg/m^3 whereas Brussels was only 18 μg/m^3; the WHO recommended limit is 60 μg/m^3. In the 1980s, Kuwait had an annual mean of 603, New Delhi 405 and Beijing of 399. Because of the obvious nature of the problems created (in the London smog of 1952 the levels of SPMs were 6000 μg/m^3 and 4700 excess deaths occurred) and the existence of technology to reduce the levels of emissions, there has been a decline in levels in most DCs in the last two decades, usually brought about by national legislation accompanied by an enforcement agency. Nevertheless, total global atmospheric concentrations of particulates seem to be increasing, though there is no agreement on what effects may follow, save that after a nuclear war, the smoke concentrations might be such as to bring about the 'nuclear winter' discussed later in the book.

A great cocktail of *gases* is emitted from the surface of the planet into its atmosphere, some from natural sources and some from human activities. As with a number of environmental contaminants, there are natural flows of many of these substances (carbon dioxide, taken up by plants during photosynthesis and emitted by animals during respiration, is an obvious case) and some of these are added to by human activities. In the case of gases, the production and use of fuels (especially the fossil fuels) stands out as a polluter at all scales. There are however some unexpected sources: domestic cattle for example emit large quantities of methane and this contributes to the possibility of global warming through the 'Greenhouse Effect' discussed on pp 183–5. Some of the airborne toxins exert no effect on the atmosphere as such but are dangerous to life or materials when they fall out.

Our present concern is focused on two major hazards: those of human health and of unpredictable global climatic change. But a number of substances produce other effects which we ought not to overlook. Some may reduce the level of amenity for instance: photochemical smog in which ozone and PAN are major constituents reduces visibility considerably, with a brown haze of varying density spreading characteristically across cities that lie in basins like Los Angeles, Tokyo or Mexico City. Such smogs may often produce no more than irritation of mucous membranes for the majority of people but thus contribute to the lowering of the quality of life. Ground-level ozone causes respiratory problems in humans and in 1988, New York City violated the federal health standard over

34 days during the summer; in Los Angeles the standard was exceeded on 172 days in the year. Mexico City topped its not-too-strict standard on 300 days in the average year. Involved in such perceptions of quality, too, are areas of forests which are suffering die-back and death from two airborne sources: sulphur dioxide and nitrous oxides (Fig 2.29). The sulphur, which rains out as part of the phenomenon of acid precipitation is accompanied in many areas by high concentrations of nitrogen oxides, mostly from car exhausts. Together, these chemicals kill or inhibit the growth of trees, especially in temperate industrial nations and reduce the economic yield of the forests as well as removing them as a source of recreation and aesthetic pleasure. In Europe, estimates for the late 1980s suggest that 71% of the forest area of Czechoslovakia is affected, 64% of Greece and the UK, and 52% of West Germany. Conspicuously clean countries like Austria still experience damage at 29%; Portugal holds up the table at only 4%.

The spatial spread of airborne gaseous pollutants is much as might be expected in the sense that many produce their greatest impact near their source, like most other substances. So carbon monoxide is only a problem in cities where it is confined by the pattern of streets and buildings or by being emitted in tunnels or underground car parks. The confining dimension is also critical in the case of cigarette smoke being inflicted on other smokers and non-smokers alike, though CO is only a transient toxic in this case, unlike the other constituents of the smoke which can affect cardiovascular and pulmonary systems. Photochemical smog, too, is worst in cities (more than a billion people are now breathing

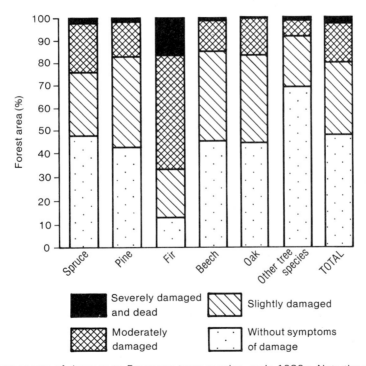

Fig. 2.29 Components of damage to European trees species, early 1980s. Note the ravages caused to firs. F. Pearce, 'The strange death of Europe's trees' *New Scientist* 112 (1537), 4 Dec 1986.

air so polluted that it breaches international safety limits) but can drift out from them to affect forests and crops in a wide plume, typically 30–80 km outside the generating agglomeration. In the USA, a total crop loss of $5.4 billion per annum is estimated from this cause. Alfalfa is especially vulnerable to ozone fallout and the US crop is diminished by 30% from ozone contamination. But most of these instances are capable of solution by technological means, such as precipitating out sulphur at power stations or installing catalytic coverters in petrol-burning vehicles. What is immensely more difficult is the tackling of those gases which have significance for global atmospheric processes since once emplaced in the stratosphere they cannot be recovered and so must be dealt with at source.

Lastly, we may mention if only in passing that indoor air pollution can be significant for humans since rooms can build up high levels of e.g. tobacco smoke, carbon dioxide and monoxide, radon (if the ground rock is a high emitter), pheromones, cosmetics and general hot air. Death is uncommon (unless deliberately self-induced) but distress and impairment of intellectual and motor skills is well documented. 'Parties pollute' is not a bad, if to some unpopular, slogan.

When fuels are burned for any purpose, waste *heat* is emitted. This may be led off directly into the atmosphere or via cooling systems such as ponds or towers. The major effects are local in scale; downwind from e.g. large power stations it is possible to detect increased incidences of cumulus cloud and of fog formation. Agglomerations like cities give off more heat since they are concentrations of energy use and their structures often absorb solar radiation during the day and give it off at night. So night-time 'heat islands' are found during still air periods. These are commonly 1–2°C above the surrounding rural areas but may be as high as 5–10°C. All the time, however, a city is radiating heat into the atmosphere and the flux may reach high levels: in London the outgoing heat flux has been measured in some districts at $100 \, w/m^{-2}$ and up to 234 in one area; the average solar input figure for London in $106 \, w/m^{-2}$. In New York, outgoing fluxes of $630 \, w/m^{-2}$ have been recorded. The extra heat of cities and the increased roughness of their surfaces seem to produce increased precipitation and thunderstorm frequency downwind, and the energy parks sometimes proposed for industrial areas might well do the same.

There are no detectable regional effects of heat emissions but the data for cities have led to some speculation as to whether heat might be a factor in any global scale warming. One set of calculations suggests that anthropogenetic emissions are at present only 0.01% of solar flux at the earth's surface, representing an energy consumption from all sources of 7.4 TW/yr. We can forget this for a while although reminding ourselves that any heat generated from fossil and nuclear materials is additional to, and does not replace, that from the sun.

The same notion, of adding to rather than replacing, natural emissions is also true of radioactivity. If we take the yearly global dose to mankind of natural radiation as 100 units, then the equivalent for medical X-rays is 19.2; for weapons testing in the atmosphere averaged over 1951–81, 13.7 units; for 1 year of civilian nuclear power 0.27 units; and 1 year of production of electricity from coal, 0.019 units. So much additional radiation is small in comparison to natural sources but is always additional to it and some of the isotopes produced have long half-lives.

Natural radiation is partly cosmic and partly terrestrial and certain human activities can increase our inescapable exposure: high-altitude flying, for example, buildings with a high radium content in the materials, or water from deep wells with a high radon content. Some activities mobilise radioactive contents, as with the burning of coal and so power station emissions inevitably have a radioactive content. Most attention is focused on technological developments such as nuclear weapons testing in the atmosphere, civil nuclear power generation in the atmosphere, and accidents. In fact, medical and dental X-rays are the greatest source of exposure for most people here, a dosage likely to grow as western medicine is extended in developing countries. Testing in the atmosphere mostly ended with the Test Ban Treaty of 1963, but the 458 atmospheric tests conducted between 1945 and 1985 produced some very high levels of fall-out of isotopes taken up by the skeleton, such as strontium-90 and caesium-137 with half-lives of 30 years and atmospheric residence times of a year or more. Fall-out was about 50% above the global average in the northern hemisphere and especially significant in tundra areas.

Nuclear power generation emits radioactive wastes to the air and the bulk of the fall-out is near the plant, though there is always controversy over whether there are any detectable effects on human health: there is even difference between most regulatory authorities and NGOs about the efficacy of the monitoring patterns. Nuclear power generation is rarely set against other systems in such calculations (Table 2.30). Accidents are few but can lead to very high levels of radiation being emitted. The partial melt-down at Chernobyl in the USSR in 1986 gave a cloud which initially contained 287×10^6 curies ($=1.06^{19}$ Bq) of radiation with a half-life over 1 day (Plate 2.32). This was 100 times worse than the largest previous accident, at Windscale in England in 1967. The Chernobyl plume drifted over Europe for 7–10 days, exposing some 400 million people to high levels but with a very patchy fall-out, which confirms that conventional ground-level monitoring may have important lacunae. But even in England and Wales, there was a 3–4% increase in background radiation and in total health-threatening levels of radiation were reached in as many as 20 countries (Fig 2.30). In the USSR there were 31 immediate deaths. To this total may be added extra cancer deaths outside the USSR: 24,000 in the next 50 years is one estimate. In such cases the atmosphere is simply a transport system: it does not process the wastes in any way, unlike water bodies which can at least provide silt which may sequester radioactive particles out of biological pathways while their activity declines.

If we can draw any general conclusions about contaminants of the air, then it is that any scale can be affected, from that of a room to the entire globe; simply because the atmosphere is a more complete unitary envelope than any other dynamic global system.

Scales of inquination and treatment

'Inquination' is a rather fancy word for pollution but it seems a pity to leave it laying unused in the dictionary, so let us use it from time to time for a change. What is obvious is that

Table 2.30 Health effects of different fuel cycles, per 1 GW/yr(e)

| Fuel Cycle | Workers | | Population | | Total | |
	Deaths	Diseases	Deaths	Diseases	Deaths	Diseases
Coal	0.1	1.5	3	1000	3	1000 *
Oil	–	0.01	3	1000	3	1000 *
Uranium	0.2	0.2	0.1	0.1	0.3	0.3
Gas	0	0	0	0	0	0

In lines with a *, best estimates are used, with uncertainty factors from 2–20. These data do not include accident hazards, in which coal outweighs uranium by a factor of 11 for fatal accidents.

Source: K. Cohen, 'Nuclear power' in Simon and Kahn (eds) 1984, 387–414.

Plate 2.32   After the Chernobyl incident, many Arctic reindeer had to be slaughtered and buried since their flesh carried too high a burden of radionuclides for human consumption. Not even the antlers could be sold to tourists, it seems.

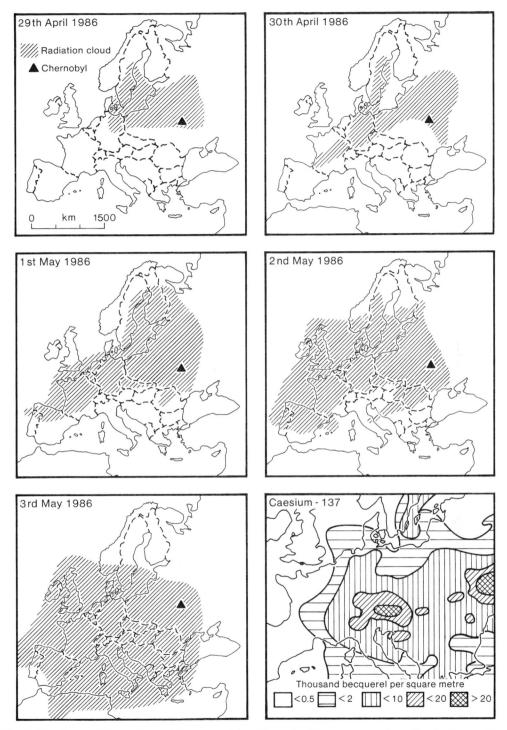

Fig. 2.30 Parts of the pattern of European fallout from the Chernobyl accident. The bottom right panel shows the overall fallout of one radionuclide – caesium-137. C. C. Park, *Chernobyl. The long shadow*. London: Routledge, 1989.

the types of inquination described in the preceding pages have different scales of impact. Thus some, while widespread in occurence, have only a minimal environmental impact and might be easy to deal with (like urban litter for instance), others are truly global in scale and impact and hence very hard to manage, as with the 'greenhouse gases'. We will examine one or two examples (but not the whole spread) of each in what follows.

The local scale

The most obvious example of this scale is the accumulation of solid wastes. Once dominated by ash from fires in the West, municipal wastes now contain plastics, paper, organic material, glass and metals. A particular set of materials is not collected by the users but dumped on the streets as litter: paper and plastics (including fast-food containers) are dominant here. Of these materials only paper is at present reasonably biodegradable, though various treatments often reduce that capacity, and genuinely degradable plastics are now beginning to enter commerce. Street and countryside litter may be unsightly but has very little ecological significance; when collected for disposal it may however be more so. It is usually disposed of into a hole and covered with earth perhaps twice per day, so that a refuse disposal site ('sanitary landfill' in North America) provides relatively little habitat for most bacteria and for rats, both of which abound in badly managed sites. However, anoxic bacterial will produce methane and this can catch fire; more modern control leads it off into the air or burns it as a source of energy. The site cannot be isolated from rainfall, of course, and leaching through it may transfer toxic materials in solution to the runoff. Good management therefore tries to make the bottom of the hole impermeable or nearly so; water must then move sideways and pass through soil or rock which adsorbs any toxic elements before joining the regional runoff. Because sites are increasingly difficult to find and manage (as standards improve), they are sometimes reserved for particular categories of especially toxic wastes (especially those produced by industry) and incineration is used for municipal materials: this latter process, for example, can also yield energy for district heating schemes.

Lead is another example of a pollutant which, while very widespread indeed in the globe (because of its transport in aerosol form), is mostly concentrated near highways as it falls out from the exhausts of petrol-driven vehicles. The only way to reduce the levels of lead (which can be especially dangerous to children) is to remove it from the fuel and many industrial nations are now taxing motor fuels in a way which encourages the use of lead-free gasoline.

Disposal of highly toxic wastes is usually via high-temperature incineration. Only a few countries possess this technology and in recent years the spectacle of ships loaded with toxics shuttling about looking for a country willing to receive the wastes has been unedifying. There is a UN Treaty on exports but its effectiveness is low (Table 2.31).

The regional scale

A major phenomenon in many places in the world but without (so far as we know) any synergistic global impacts or consequences is the eutrophication of fresh waters (p. 162–3). The causes of this can be agricultural (runoff from excessive fertiliser use), domestic (high concentrations of untreated sewage) or urban-industrial (effluent containing elements like phosphorus) but the end results are similar. The perception of the inquination may often be sporadic because the inputs are pulsed, as when heavy rain occurs soon after a farmer has applied very heavy doses of nitrogenous fertiliser to bare soils or when dry weather concentrates sewage in rivers with low flows and warm water. Whatever the cause, the effects

Table 2.31   Toxic trade: Hazardous and special waste

	Production	1000 tonnes Imports	Exports
USA 1985	25,000	40	150
W. Germany 1985	5,000	75	700
Britain 1986	3,900	83	–
Canada 1980	3,290	120	40
France 1987	2,000	250	25
Holland 1986	1,500	–	155
Sweden 1980	500	–	15

Source: OECD

are much the same and so is the remedy, for there is no known technology for removing nitrogen and phosphorus once in solution, from free water-courses. (Though vegetation will do some scavenging.) It is imperative that this problem is tackled at source either by lowering the levels of N and P used in various processes (e.g. in detergent usage) or by pre-release treatment. Sewage treatment of the usual kind, for example, removes solids and bacteria (visit any modern installation and your guide will drink a cup of the effluent as it leaves the plant) but not N and P in solution. So in the 1980s the North Sea received 100,000 million t/yr of phosphorus from rivers, increasing its phosphate load by sevenfold. Algal blooms thus released may emit a compound which changes to sulphuric acid in the atmosphere and adds to the acid precipitation problems). Treatments exist but it is often the judgement that money is better invested in extending ordinary treatment to sewage now left raw, than removing the generally less offensive nitrogen and phosphorus. So in most nations, eutrophication of fresh waters seems likely to continue.

A regional sink is a description which might be applied to enclosed sections of the world ocean like the North Sea. In spite of some international agreements, the basin receives annually some 5.5×10^6 tonnes of industrial waste, 5000 t of sewage sludge, 97×10^6 t of dredged material together with the emissions from 94,000 t of industrial waste incinerated at sea, a practice to stop in 1994. About 25,000 t of hydrocarbons were either spilled or discharged into the sea, and radionuclides originally from the Irish Sea amounted to 2000 TBq. Britain is the only country still dumping sewage sludge (though it has promised to stop this) and the sludge has yearly contained 0.5 t of mercury and 3.0 t of cadmium. Algal blooms ('red tides'), seal disease epidemics and fish diseases affecting up to 30% of flatfish are all ascribed by some to contamination. The British government asserts that the North Sea is, overall, a healthy body of water. The Dutch, on the other side, have referred to it as one of the most polluted seas on earth.

As suggested earlier in this chapter, acid precipitation is also a good example of a regional phenomenon. Maps (e.g. Fig 2.31) can be produced of regional pH values which demonstrate that certain areas experience acidification of both dry and wet deposition from the atmosphere. Historic buildings, as in Krakow in Poland, Athens in Greece, and Prague in Czechoslovakia, are often very visibly damaged. Sudden acid mists can defoliate trees and corrode, metals over wide areas and may have pH values as low as 2.0, which is more acid than vinegar. This then affects most systems, although fresh waters and forests have received the most attention. The complexities of the chemistry are considerable: compounds of both sulphur and nitrogen are implicated, for instance, but some phenomena such as tree die-back can be caused either by acid precipitation originating in power station emissions hundreds of kilometres away (there are instances of sulphur dioxide being transported 1000–2000 km in a period of 3–5 days) or by nitrogen oxides from the exhausts of vehicles bringing family groups to picnic. Thus the management of acid deposition can be quite complex, though mopping up by e.g. liming lakes is only for a few places. At-source control involves firstly removing sulphur from emission sources like power stations which burn high-sulphur coals and oils. Then, complicated converters need to be applied to vehicle exhaust systems to convert e.g. NO_x compunds to less reactive substances. One problem here is that converters tend to reduce performance so that drivers may tend to put their foot down harder and hence increase CO_2 output from the extra fuel consumed. At that level, a regional problem interacts with the global scale.

A different type of regional problem has come to light with the demise of the highly secretive hard-line communist governments of Eastern Europe. It has now become plain that little attention to pollution control was paid in these nations and so their land, water bodies and atmosphere are highly contaminated. Industries have been the major source of this: in Poland even the old government declared danger zones and disaster areas: five of the latter are in Upper Silesia, where sulphur dioxide deposits exceed 100 tonnes per km^2/yr. In Czechoslovakia, 70% of the rivers are polluted by mining wastes; in Romania, the old regime built 4500 water purification plants, none of which work. The former East Germany's citizens had the notable distinction of emitting more SO_2 per head than anywhere else in the world. The task of bringing these areas to reasonable Western standards is formidable, let alone improving them to meet the newer internationally-agreed standards which seem inevitable.

Many environmental systems were affected by the oil spills and well-burning which accompanied the Gulf War of 1991. More than 500 oil wells were set on fire and 3–6 million barrels a day of oil consumed, producing an estimated 500,000 t/month of soot which in March 1991 formed a cloud about 8 million km^2 in size, reducing the temperature underneath it by some $10°C$. In addition, perhaps 4 million t/yr of sulphur will add to nitrogen oxides in the smoke to produce a very acid rain. The rain is in fact black, as it was at Hiroshima. The overall effects of this terrible waste (10% of Kuwait's oil reserves may be lost before the fires are put out) are likely to be regional rather than global but will be upon nations (Turkey, Syria, Iraq, Jordan, Iran) least able to tolerate the impending reductions in agricultural production. Food production, too, seems to be the most important consequence of the oil spilled into Gulf waters. Estimates of 3–7 million barrels are made: there is certainly more than the *Exxon Valdez*. Photosynthesis and larval habitats like salt-marshes and coral reefs are affected as well as the fish populations themselves, which may be either killed or tainted.

The global scale

Two giant systems have the potential to affect the biogeochemical functioning of any other part of the globe since they form continuous pools: the atmosphere and the oceans. For many decades it was thought that they were so vast that no human activity could possibly change their nature but in the last 20 years scientific evidence has been presented to firstly a sceptical and then an awed public and their politicians, that mankind could indeed promulgate detectable alterations with potentially serious consequences.

The current focus for most work is the atmosphere. Two processes have been identified which alter the function and parameter values of the atmosphere to an extent which threatens human life and socio-economic systems. The first of these is the depletion of the ozone (O_3) layer of the stratosphere which is about 16–25 km above the planetary surface. The layer absorbs solar radiation in the ultraviolet B wavelengths and also promotes a

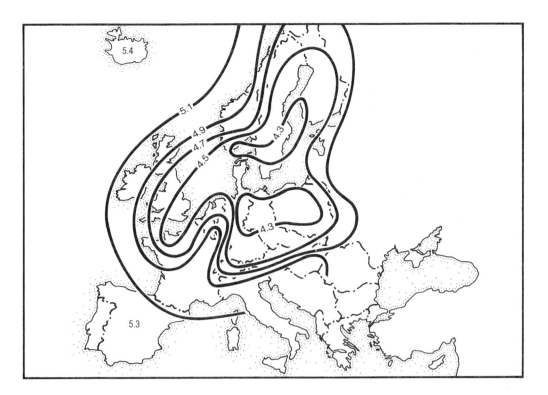

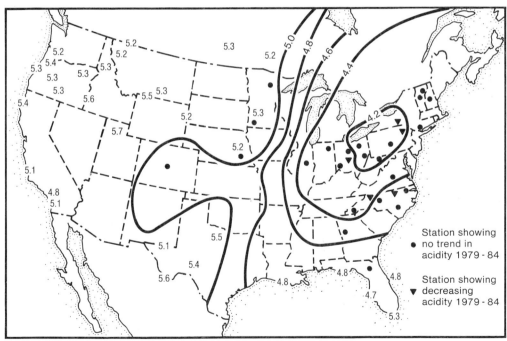

Fig. 2.31 The distribution of acidity in the precipitation of Europe and North America in 1985, in pH units. Note in each case the existence of a core area of very low pH which equals very high acidity. *World resources 1988–89*. NY: Basic Books Inc., 1988.

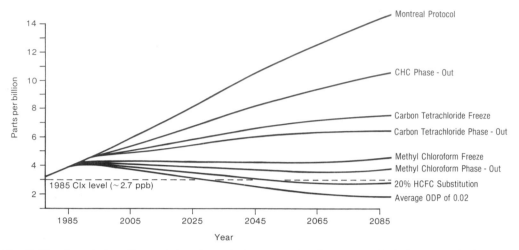

Fig. 2.32 Cumulative effects of the planned phase-out of various ozone-depleting gases in the atmosphere: some possible scenarios. Clx is the totality of the chlorine-bearing compounds; ODP = Ozone Depleting potential. (Friends of the Earth)

temperature inversion between 15 and 50 km which as it were 'contains' the convective processes of the troposphere which produce our weather and climate. Any attenuation of the ozone layer would then affect meteorological processes, presumably in an unpredictable way, and increased UV-B transmission may produce some global warming, though the exact chain of causality (involving models of albedo, aerosol hazes and cloud formation) is not diamond clear. Higher levels of UV-B will also lead to enhanced rates of human skin cancer.

For some years now, evidence has been accumulating that the ozone layer has been thinning at a rate which has added up to a loss of up to 3% in the last 20 years, and that in the spring the polar regions (especially the Antarctic south of 60°) exhibit 'holes' where the O_3 levels are reduced by 5% or more. In 1990, the area of reduced ozone over the Antarctic was about 14×10^6 km² in area.* The concentration of ozone depends upon interactions with very low levels (of parts per billion or trillion, usually called 'trace gases') of oxygen, oxides of nitrogen and hydrogen, and chlorine, and it seems to be a fact that human activity has increased the quantities of these trace gases in the stratosphere (Table 2.32), where their residence times enable them to react with ozone and deplete the 'ozone layer'. The main identifiable source of depletion have been the group of chemicals called chlorofluorocarbons (CFCs) which are used in aerosol propellants, refrigeration, foam-blowing and as solvents. Global production rates are not accurately known but emissions are estimated to have been 100 t in 1931, 35,000 t in 1950 and 649,000 t in 1985. Implication of these chemicals has led to the Montreal Protocol of 1987 which aims to freeze CFC production at 1986 levels (substitutes are available in most consumer products but less easily for cheap refrigerator coolants in LDCs) and to reduce emissions to 50% of 1986 levels by the end of this century (Fig 2.32). Even with a fully-ratified Montreal Protocol, chlorine levels in the atmosphere seem likely to rise from 3 ppb (1989 level) to 6–9 ppb in the next few decades. The likely effectiveness of international measures currently contemplated is complicated by the role of other ozone-depletors such as carbon tetrachloride and methyl chloroform, which are not explicitly mentioned in the Protocol.

*It seems to be spreading to temperate areas as well.

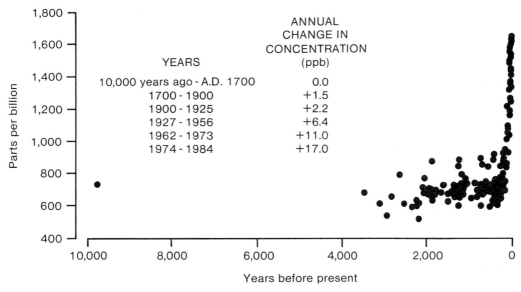

Fig. 2.33 Annual changes in methane gas over the last 10,000 years in parts per billion (ppb). *World resources 1988–89.* NY: Basic Books, 1989 [Fig. 23.1].

Table 2.32 The 'Greenhouse' gases

Gas	Uses/ production	% Contribution to global warming
Carbon dioxide (CO_2)	Combustion of fossil fuels (80%; Biomass burning (20%)	50
Methane (CH_4)	Biomass burning, anerobic fermentation in wetlands and padi, cattle guts; natural gas exploration and transport	18
Chlorofluoracarbons (CFC's) and halons	Aerosol pollutants, coolants, foamers Fire extinguishers	14
Low level ozone (O_3)	Formed by action of sunlight on hydrocarbons and nitrogen oxides from vehicles	12
Nitrous oxide (N_2O)	Fertilizers, combustion of fossil fuels, changing land use	6

Compiled from a number of sources

CFCs share with many other gases, both natural and human-produced, the ability to be relatively 'transparent' to incoming solar radiation but to inhibit the loss to space of outgoing radiation. The heat balance of the troposphere depends upon the balance of these two fluxes and so gases which retard outgoing radiation trap heat, leading to the so-called 'greenhouse effect'. As with the ozone layer a number of gases have been identified (Table

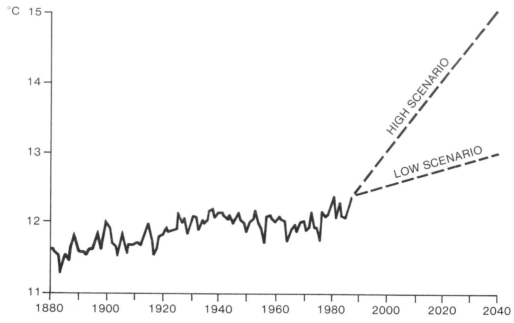

Fig. 2.34   Trends in observed global temperatures in the 20th century, two scenarios for future years. H. F. French, *Clearing the Air*. Washington DC: Worldwatch Institute Paper 94, 1990.

2.32) whose concentration in the atmosphere has been affected by human activity (Fig 2.33) and which are implicated in a global warming which has been measured for the oceans at $0.6 \pm 0.3°C$ in the last 90 years.

If emissions continue at their present rate (usually labelled the 'Business-as-Usual' scenario, then predictions are for a global rise in temperature in the next decade of about $0.3°C$ per decade, with an uncertainty range from $0.2-0.5°C$. (Fig 2.34) By 2025 this would produce an increase of $1°C$ above the present level (and hence about $2°C$ above the pre-industrial level) and $3°C$ above the present level before the end of the next century. Such changes may produce many effects upon ecosystems, agriculture, species distributions (those of pests and diseases, for example) and on sea-levels. The B-a-U scenario yields a global estimate of 6 cm per decade for the next century (with an uncertainty range of 3–10 cm) which results in a rise of about 20 cm by 2030 and 65 cm by 2099. Significant regional variations are to be expected.

The gas most strongly identified with global warming is carbon dioxide (CO_2). In the late 19th century its concentration in the atmosphere was about 275 ppm, and in the late 1980s, 352 ppm. The quantity of CO_2 in the atmosphere depends upon a complex series of interactions, and the quantitative flows are not all known. Especially important are the emissions from the combustion of fossil fuels (*ca* 5 Gt/yr, table), the contribution from burning of forests and grasslands and of the oxidation of humus, and the take-up by vegetation and the phytoplankton of the oceans (Table 2.33). The world's tropical regions are at present yielding a net carbon release of 0.6–1.1 Gt/yr. The soils and biomass of the temperate zone may be a small source or a net sink. Whatever the details, attention is mostly focused on fossil fuels as a source and on the removal of forests with high rates of primary productivity like the TMFs, which sequester carbon as wood.

Table 2.33   Annual emissions of carbon dioxide, actual and projected, $\times 10^6 t$

	1986 (% of world share)		2030		% annual growth
WORLD	5,575	(100)	18,184	(100)	2.7
USA	1,299	(23)	3,257	(18)	2.1
USSR	1,030	(18)	2,940	(16)	2.4
China	621	(11)	1,218	(7)	1.5
LDCs	1,452	(26)	5,891	(32)	3.2
Japan	260	(5)	419	(2)	1.1
S.E. Asia	146	(3)	532	(3)	3.0

Source: *The Environmental Digest* no. 27, August 1989, p. 6

The forcing of radiative warming depends, however, not only on carbon dioxide. Other gases, though smaller in quantity, may be effective heat-retainers. A molecule of methane, for example, is about 21 times more effective then carbon dioxide and CFC-11 about 12,000 times more effective. When the effectiveness of each is multiplied by the quantity emitted, the Global Warming Potential (GWP) of each can be estimated. These are usually expressed relative to 1 kg of CO_2 for a future date and are of course estimates. So for a time horizon of 20 years from 1990, if the GWP for carbon dioxide is 1, then that for methane is 63, for nitrous oxide is 270, for CFC-11 is 4500, for CFC-12 is 7100 and for HCFC-22 is 4100. The contribution of the gases also varies with the lifetime: that for CO_2 is long, whereas that for methane is quite short. So the GWP for a 100-year horizon based on the 1990 level of emissions are (again relative to $CO_2 = 1$): methane = 21; nitrous oxide = 290 and HCFC-22 = 1500. Other are too uncertain to figure in these estimates. The estimates change as new and more subtle models are developed, but this global phenomenon poses a challenge to human society as a whole. Thus the late 1980s have been marked with a number of international meetings at both scienfic and political level aimed at reducing the output of 'greenhouse' gases; carbon dioxide in particular presents difficulties because it is tied in with the energy consumption which, as we have seen, is so interwoven with the lifestyle of the industrial nations. Whatever the cost, it seems as if global carbon emissions will have to be curtailed and some jurisdictions have already set targets. Where current emissions per capita are low in LIEs (e.g. Kenya and India at <0.5t/cap/yr) then growth must be permitted. But those with emissions of over 2.0 t/cap/yr (most HIE's) need to reduce these outputs by 2.0–3.0% per year, starting now.

Waste exports, already mentioned briefly, have a set of supra-national dimensions. Industrialised nations probably ship ca 4 million t/yr of highly toxic wastes to LDCs, who will 'dispose' of them for about $40/t instead of the $250/t for burying or $1500/t for incineration in the HIEs. At least 10 African countries act as recipients: Benin for example has agreed to receive 5 million t/yr from N. America and Europe, and radioactive wastes from the USSR. An American company has offered to send the Marshall Islands 25 million t/yr of garbage over the next five years, to keep out the rising sea-levels.

A last possible global pollution would be the effects of a medium to large scale nuclear war. The modelling of the extra UV-B, the radioactive fallout over a long period and the climatic scenarios have produced the concept of the 'nuclear winter' and its milder variant, the 'nuclear autumn'. The details are not important here: what is necessary to note is that it might be even *possible* and that after it the whole of the world's ecosystems of all kinds would be totally disrupted: a kind of macrocosmic Year Zero.

Problems of priorities

All these undesired conditions manifest themselves at the same time. Since it is never possible to tackle them all at once, some kinds of priorities have to be distinguished. There is little doubt that the global futures posited by modelling based on early signs of climatic change carry the greatest suggestions of menace: whole nations of the Indian and Pacific Oceans might disappear, world food production patterns would need reorganising, and semi-natural vegetation would also probably be subject to stresses. Increased UV-B radiation carries direct threats to human health and may affect other animals and plants as well, though knowledge is as yet at an early stage. And as mentioned above, high amplitude fluctuations on the route to a new equilibrium are dangerous destabilisers for human and natural systems alike.

Thus these long-term probabilities have top priority in governmental and international councils and in the early 1990s are likely to dominate the international agenda, to the exclusion of regional and local issues. Yet there are linkages between scales. At the material level, nitrogen oxides are implicated in acid deposition as well as in 'greenhouse' phenomena, for example and so containment of these gases would help with both sets of problems. At the behavioural level, there is no doubt that individuals and NGOs need to be involved and to feel that they are playing a significant role and this is not immediately possible at the global level. Hence the local and regional issues, from litter to wildlife, while not of the same planetary importance in an objective, scientific, sense are essential if the populations that are the resource-consuming, waste-producing ones are to modify their habits and demands. Hence the ways in which human individuals and groups take cognisance of these processes is clearly important.

Viewing the process

At the heart of discussing global waste management is the question, what sort of model of the world do we have? Is it one which exists solely to serve our own interests? If so, then does it work like a snooker table? That is, can we hit the colours neatly and not disturb the whole pattern badly unless it is to place another ball ready for the next shot? Or is it perhaps more like a child's mobile in which every breath of warm air moves one piece and then all the others move by way of adjustment? These are only similies of course, and the truth lies elsewhere. But a major contribution of ecology and now quantum physics has been to lift our worldview from the snooker table to the mobile near the ceiling: beyond it there is indeed a limit.

Societies and wastes

We begin, as we must, with the individual. All of us create wastes roughly in proportion to our use of resources, so that the poor are much less responsible than the rich. Here, experience in the West is that many individuals feel much more empowered to mop up after themselves if encouraged to do so by the next-level mechanisms of society. A local government body can hence do more than it thinks by encouraging waste separation prior to collection (for cheaper recycling), for example. Voluntary organisations of concerned individuals are also essential catalysts in almost every process, for they usually provide the initial surge of energy that implants the idea of something new: governments in this field are almost all conservative.

At national level, the legal and policy frameworks of the national state become integral parts of the waste management process. There are two basic methods which may underlie a set of laws and regulations regarding wastes. These are:

- Regulatory methods, in which the civil and possibly criminal laws are formulated to prohibit the discharge of wastes to the environment except under carefully specified provisions. Thus the rule is basically, 'You may not dispose of substance X except by means of high-temperature incineration; if you do so then your firm will pay a fine of X thousand and on the second occasion the Managing Director will go to prison'. An objective and preferably incorruptible regulatory agency is needed with these methods.

- 'Polluter pays methods'. Here the regulating agency issues licences for discharge, the price of which is fixed so as to encourage the discharger to treat the waste himself. Should he wish not to do so, then the price of the licence provides for the cost of a public clean-up. In Sweden, there is a tax on energy emissions. At present this is SKr30 per kilo on sulphur and Skr0.25 on carbon. From 1991, No_x emissions will be charged at Skr40 per kilo. As with regulations, an enforcement agency is needed; in both the involvement of citizens, citizen groups and the police seem to be needed to cope with those who try to evade any set of laws or licences.

At international level, the regulatory method is paramount, proceeding by means of Protocols, Conventions and Treaties. Either a special single-function body or the UNEP is detailed to set the conditions and require their enforcement, or the appropriate governmental agency of the acceding countries provides the monitoring system. This latter is more easily swayed by national interests and the whole procedure lacks teeth in any case: only exposure in various international courts and the resulting bad publicity normally ensue. Appropriate national fines on a very large scale are not used nor is the attractive alternative of sending an offending nation's Environment Minister to prison for a while.

Worldviews of wastes

One general feature of wastes and their disposal is the way in which this 'back-end' of resource processes can be seen by a public as separate from the other parts of the flow. Clean-up is often perceived as a responsibility of individuals and governments, but the process which results in the wastes is a function of a free market or a centralised planning system. Both of these are set up to deliver materials. Thus there is a segmented approach to the resource process. An overview, which necessarily would normally be taken by a government, might for example enact collective regulations to ensure that material consumption is not thrown at people (as in some advertising campaigns) with the aim of making them feel inadequate if they do not consume at a certain level. Similarly, production ought not to be forced on people but powerless individuals and groups often are made to play host to an industrial plant (e.g. a poisonous materials) that nobody else wants. Some resource processing thus becomes simply a reflection of political power structures, with the upper groups playing NIMBY (Not In My Back Yard). Increased awareness of the links between resource processing, consumption and waste production does however give the individual some chance to act. Recycling, for example, is beneficial to the environment in all kinds of ways and can usually be run without large-scale social or economic disruption (Tables 2.34 and 2.35). Nevertheless even after the extraction of recyclable items, municipal solid wastes contained (by weight) 30% of containers and packaging and 24.5% of garden waste.

Individual action is well summarised in the 'five Rs' approach to materials' use and waste treatment, which can involve not only the individual but also local action and thus bring about a feeling of being involved in the management of change:

Table 2.34 Wastepaper use in the EEC during the early 1980s

Product	Wastepaper proportion of furnish (%)	Proportion of EEC total use of wastepaper (%)
Corrugated board	88	45
Other board	72	30
Packaging paper	50	8
Creped paper and tissue	40	5
Newsprint	26	5
Printing papers	4	3
Others	39	4

Source: K. Ebeling, 'Technological developments in the pulp and paper industry', in M. Kallos *et al* 1987

Table 2.35 Environmental benefits derived from substituting secondary materials for virgin resources

Environmental benefits	Aluminium	Steel	Paper	Glass
		(percent)		
Reduction of:				
Energy Use	90–97	47–74	23–74	4–32
Air Pollution	95	85	74	20
Water Pollution	97	76	35	–
Mining Wastes	–	97	–	80
Water Use	–	40	58	50

Compiled from various sources

- *Refusal* The reinforcement of an individual's refusal to be defined by possessions, or to have forced on them completely unnecessary items such as excessive packaging of the kind which is heavily material- and energy-consumptive. This is very hard in western nations because of commercial and peer-group pressures. It is possible to add to this item the refusal of war, the course and preparation of which uses immense amounts of resources in laying waste to everything in sight.

- *Reduction* The deliberate slowing-down of the rate of consumption of materials which produce high levels of wastes and the substitution of lower-waste items. In the West, eating less meat would be a good example, as would turning down the heating thermostat and putting on a sweater.

- *Re-use* The recognition that second-hand goods are often adequate for the purpose intended and that to buy new is often not at all necessary except to improve our image in somebody else's eyes. A used car or bicycle is often totally adequate especially if it has been built with the next criterion in mind.

- *Repair* Some goods have high levels of built-in obsolesence so that repairs are either impossible (the 'throw-away society') or so expensive that buying new is cheaper. A shift from this position is technologically possible in many cases but encouragement of the manufacturers is a complex affair: not many car makers, for example, offer a life-time guarantee with a vehicle.

- *Recycle* This involves the recognition that re-cycling of materials is often cheaper in money terms and in energy consumption than using virgin sources. Where it is not then it is usually the case that the costs of dealing with the discarded items is not included in the price, i.e. in economic terms it is an external cost. Hence the Swedish government, for instance, levies a tax on car sales which is to cover the eventual cost of recovering the materials of the car when its lifetime has expired. There is no incentive then for the last owner to leave the hulk to rust on a city street or in the countryside. At the very least, a trip to the scrapyard is called for. In Oregon, USA, all glass bottles have a deposit on them which encourages their return to a shop for recycling rather than relying on special trips by the concerned to a bottle bin; this latter is probably more energy-consumptive than simply taking bottles back when buying another case of Ch. Latour Grand Cru 1976. It lacks however the psychic satisfaction of hearing bottles smash as they are lobbed through the holes.

A last shift in our behaviour is to learn from our own past and from the present of poorer societies: to treat waste accumulations as mines of recoverable materials (Plate 2.33). In most wastes there are substances which can be recycled or burned to provide energy. Only economics prevents the development of advanced technology which would sort all these

Plate 2.33 Solid wastes and disposal problem for municipalities as in this Moroccan example. These children scavenge aluminium and re-sell it and are thus in the vanguard of a growing movement to 'mine' such wastes for materials and energy. Whether the boys see it like that is another matter.

items for further use and economics is an artefact of society and not a manifestation of something ineluctable like the law of gravity. In other words, any consideration of resource use must be cognisant of the whole resource process, from extraction through to the disposal of the wastes. Any approach to waste treatment should look forward to the resource use and ask why the waste was generated in the first place. Here then is a chance to link the segments of the resource process together and in so doing mimic what millenia of evolution have provided for natural ecosystems.

Further reading

Energy: the binding resource

Aitchison, J. and Heal, D. 1990: 'Which fuel? Diversity in world energy consumption', *Geography Review* 3 (5) 18-21.
Basalla, G. 1982: 'Some persistent energy myths', in Daniels, G. H. and Rose, M. H. (eds) *Energy and transport: historical perspectives on policy issues*. Beverley Hills and London: Sage, 27-38.
Chandler, W. U. 1985: Energy productivity: key to environmental protection and economic progress. Washington DC: *Worldwatch Paper* 63.
Cook, E. 1976: *Man, energy, society*. San Francisco: Freeman.
Diamond, J. R. 1987: 'Human use of world resources', *Nature* 328, 479-80.
Flavin, C. 1990: 'Slowing global warming', in Brown, L. R. (ed) *State of the world 1990*. New York and London: W. W., Norton 17-38.
Flavin C. and Durning, A. B. 1988: Building on success: the age of energy efficiency. Washington DC: *Worldwatch Paper* 82.
Hall, C. A. S. Cleveland, C. J. and Kaufmann, R. 1986: *Energy and resource quality. The ecology of the resource process*. New York: Wiley.
Lonard, A. and Lonard, E. D. 1983: 'Evaluation of the mutagenic potential of different forms of energy production', *The Science of the Total Environment* 29, 195-211.
Pimentel, D. *et al* 1984: 'Environmental and social costs of biomass energy', *BioScience* 34, 89-94.
Smil, V. 1987: *Energy food environment. Realities myths options*. Oxford: Clarendon Press.

Food and agriculture

Brown, L. R. and Young, J. E. 1990: 'Feeding the world in the Nineties', in L. R. Brown (ed) *State of the world 1990*. London and New York: W. W. Norton, 59-78.
Conway, G. R. and Barbier, E. B. 1990: *After the green revolution. Sustainable development for agriculture*. London: Earthscan Publications.
P. Hatchwell (1989) The risks of releasing genetically engineered organisms *The Ecologist* 19 (4) 130-36.
Dover, M. and Talbot, L. M. 1987: *To feed the earth: agro-ecology for sustainable development*. Washington DC: World Resources Institute.
Duckham, A. N. Jones, J. G. W. and Roberts, E. H. 1976: *Food production and consumption. The efficiency of human food chains and nutrient cycles*. Amsterdam: North-Holland.
Knorr, D. (ed) 1983: *Sustainable food systems*. Westport, Conn: Avi Publishing Co.

Langley, J.A. Heady, E.O. and Olson, K.D. 1984: 'The macro implications of a complete transformation of U.S. agricultural production to organic farming practices', *Agriculture, Ecosystems and Environment* 10, 323–33.

Lapp, F.M. and Collins, J. 1989: *World hunger: 12 myths.* London: Earthscan Publications.

Pearse, A. 1980: *Seeds of plenty, seeds of want. Social and economic implications of the Green Revolution.* Oxford: Clarendon Press.

Stanhill, G. 1990: 'The comparative productivity of organic agriculture', *Agriculture, Ecosystems and Environment'* 30, 1–26.

Smil, V. 1987: *Energy, food, environment. Realities, myths, options.* Oxford: Clarendon Press.

Tarrant, J.R. 1980: *Food policies.* Chichester; Wiley.

Tudge, C. 1988: *Food crops for the future.* Oxford: Blackwell.

Woods, R.G. (ed) 1981: *Future dimensions of world food and population.* Boulder, CO: Westview Press.

Buckwell, A. and McKey, A. 1990: 'Biotechnology and agriculture', *Food Policy* **15**, 44–56.

Forestry

Apin, T. 1987: 'The Sarawak timber blockade', *The Ecologist* 17, 186–88.

Earl, D.E. 1975: *Forest energy and economic development.* Oxford: Clarendon Press.

Ebeling, K. 1987: 'Technological developments in the pulp and paper industry', in M. Kallio *et al* (eds) *The global forest sector: an analytical perspective.* Chichester: Wiley, 224–52.

Flavin, C. and Pollock, C. 1985: 'Harnessing renewable energy', in L.R. Brown (ed) *State of the world 1985.* NY and London: W.W. Norton & Company.

Goldman, M.I. 1972: *The spoils of progress: environmental pollution in the Soviet Union.* Cambridge, Mass: MIT Press.

Hepburn, G. 1989: 'Pesticides and drugs from the Neem tree', *The Ecologist* 19, 31–32.

Hodgson, G. and Dixon, J.A. 1989: 'Logging versus fisheries in the Phillipines', *The Ecologist* 19 (4) 139–43.

Ives, J.D. and Messerli, B. 1989: *The Himalayan dilemma. Reconciling development and conservation.* London and NY: Routledge/UNU.

Jordan, C.F. 1985: *Nutrient cycling in tropical forest ecosystems.* Chichester: Wiley.

Jordan, C.F. 1987: 'Conclusion', in C.F. Jordan (ed) *Amazonian rain forests. Ecosystem disturbance and recovery.* New York: Springer-Verlag, Ecological Studies vol 60, 100–21.

Kallio, M. Dykstra, D.P. and Binkley, C.S. (eds) 1987: *The Global forest sector. An analytical perspective.* Chichester: Wiley for IIASA.

Likens, G.E. *et al* 1977: *Biogeochemistry of a forested ecosystem.* New York: Springer-Verlag.

Mann L.K. *et al* 1988: 'Effects of whole-tree and stem-only clearcutting on postharvest hydrologic losses, nutrient capital and regrowth', *Forest Science* 34, 412–28.

Fuelwood

Pearce, G. 1990: 'Fruits of the rainforest', *New Scientist* 125, 42–45.

Postel, S. and Heise, L. 1988: Reforesting the earth. Washington DC: *Worldwatch Paper* 83.

Pryde, P.R. 1972: *Conservation in the Soviet Union.* CUP.

Repetto, R. and Gillis, M. (eds) (1988) *Public policies and the misuse of forest resources.* Cambridge University Press for WRI.

Shea, C.P. 1988: Renewable energy: today's contribution, tomorrow's promise. Washington DC: *Worldwatch Paper* 81.

Siebert, S.F. 1987: 'Land use intensification on tropical uplands: effects on vegetation, soil fertility and erosion', *Forest Ecology and Management* 21, 37–56.

The Environmental Defense Fund 1987: 'The failure of social forestry in Karnatka', *The Ecologist* 17, 151–54.

Totman, C. 1989: *The Green archipelago. Forestry in pre-industrial Japan.* Berkeley and Los Angeles: University of California Press.

Young, A. 1989: *Agroforestry for Soil Conservation.* Wallingford, UK: CAB International.

Eckholm E. 1979: 'Planting for the future: forestry for human needs'. Washington DC: *Worldwatch Paper* 26.

Water

Biswas, A.K. (ed) 1978: *United Nations water conference. Summary and main documents.* Oxford: Pergamon Press, Water Development, Supply and Management, vol 2.

Gower, A.M. (ed) 1980: *Water quality in catchment ecosystems.* Chichester: Wiley.

Hollis, G.E. (ed) 1979: *Man's impact on the hydrological cycle in the United Kingdom.* Norwich: Geobooks.

Postel, S. 1985: Conserving water: the untapped alternative Washington DC. *Worldwatch Paper* 67.

Rogers, P.R. 1985: 'Fresh water', in R. Repetto (ed) *The global possible.* New Haven and London: Yale UP, 255–98.

Shiklomanov, I.A. 1990. 'Global water resources', *Nature and Resources* **26**(3), 34–43.

Speidel, D.H. Ruedisili, L.C. and Agnew, A.F. (eds) 1988: *Perspectives on water. Uses and abuses.* New York and Oxford: Oxford University Press.

White, G.F. 1984: 'Water resource adequacy: illusion and reality', in J.L. Simon and H. Kahn (eds) *The resourceful earth.* Oxford: Blackwell, 250–66.

Wolman, M.G. and Wolman, A. 1986: 'Water supply: persistent myths and recurrent issues', in I. Burton and R. Kates (eds) (eds) *Geography, resources and environment. Vol II. Themes from the work of Gilbert F. White.* Chicago and London: University of Chicago Press, 1–27.

The Oceans: renewable resources

Barnes, R.S.K. and Mann, K.H. 1980: *Fundamentals of aquatic ecosystems.* Oxford: Blackwell Scientific Publications.

Cushing, D.H. 1988: *The provident sea.* CUP.

Garcia, S. Gulland, J.A. and Miles, E. 1986: 'The new law of the sea and the access to surplus fish resources', *Marine Policy* 10, 192–200.

Holt, S. 1985: 'Whale mining, whale saving', *Marine Policy* 9, 142–213.

Klee, G.A. 1980: Oceania, in G.A. Klee (ed) *World systems of traditional resource management.* London: Edward Arnold, 245–81.

Larkin, P.A. 1977: 'An epitaph for the concept of maximum sustainable yield', *Trans Amer Fish Soc* 106, 1–15.

Mason, C.M. (ed) 1979: *The effective management of resources. The international politics of the North Sea.* London: Frances Pinter Ltd.

Pearce, F. 1991: 'A dammed fine mess', *New Scientist* 130, 36–39.

Pitcher, T. J. and Hart, P. J.B. 1982: *Fisheries ecology.* London and Sydney: Croom Helm.

Ryther, J.H. 1969: 'Photosynthesis and fish production in the sea', *Science* 166, 72–76.

Steele, J.H. (ed) 1970: *Marine food chains* Edinburgh: Oliver and Boyd.

Non-renewable resources

Arndt, P. and Lüttig, G. W. (eds) 1987 *Mineral resources extraction, environmental protection and land-use planning in the industrial and developing countries.* Stuttgart: E. Schweizerbart'sche Verlagsbuchhandlung.

Dasgupta, P. 1989: 'Exhaustible resources', in L.E. Friday and R. A. Laskey (eds) *The fragile environment.* CUP, 107–26.

Flawn, P. 1966: *Mineral resources.* Chicago: Rand McNally.

Goeller, H. E. and Weinberg, A. M. 1976: 'The age of substitutability', *Science* 191, 683–9.

Govett, G. J. S. and Govett, M. H. (eds) 1976 *World mineral supplies. Assessment and perspective.* Amsterdam: Elsevier Developments in Economic Geology, 3.

Rees, J. 1989: *Natural resources. Allocation, economics and policy.* London: Routledge, 2nd edn.

Rosenberg, N. 1980: 'Technology, natural resources and economic growth', in C. J. Bliss and M. Boserup (eds) *Economic growth and resources.* Vol 3. *Natural resources.* London: Macmillan.

Environments as resources

Anderson, I. MacKenzie, D. and Dickson, D. 1989: 'Should Antarctica be left on ice?', *New Scientist* 124, 28–29.

Bonner, W. N. and Walton, D. W. H. (eds) 1985: *Antarctica.* Oxford: Pergamon Press Key Environments series.

Comito, T. 1978: *The idea of the garden in the renaissance.* Hassocks: Harvester Press.

Cross, M. 1991: 'Antarctica: exploration or exploitation?', *New Scientist* 130, 29–32.

Harvey, J. 1981: *Mediaeval gardens.* London: Batsford.

Hudson, R. J. Drew, K. R. and Baskin, L.M. (eds) 1989: *Wildlife production systems. Economic utilisation of wild ungulates.* CUP.

IUCN/WWF/UNEP 1980: *World conservation strategy.* Gland, Switzerland: IUCN.

Lucas, R. C. 1989: 'A look at wilderness use and users in transition', *Natural Resources Journal* 29, 41–55.

Mathieson, A. and Wall, G. 1982: *Tourism: economic, physical and social impacts.* London and New York: Longman.

McCloskey, J.M. and Spalding, H. 1989: 'A reconnaissance-level inventory of the amount of wilderness remaining in the world', *Ambio* 18 (4) 221–27.

McNeely, J. A. and Miller, K. R. (eds) 1984: *National Parks, conservation and development. The role of protected areas in sustaining society.* Washington DC: Smithsonian Institution Press.

Myers, N. (ed) 1985 *The Gaia atlas of planet management.* London and Sydney: Pan Books, 152–169.

OECD 1987: *Tourism policy and international tourism in OECD member countries.* Paris: OECD.

Parsons, Sir A. (ed) 1987: *Antarctica: the next decade.* CUP.

Pearce, D. 1987: *Tourism today. A geographical analysis.* London: Longman.

Sattaur, O. 1989: 'The shrinking gene pool', *New Scientist* 123, 37–41.

Shapley, D. 1985: *The seventh continent. Antarctica in a resource age.* Washington DC: Resources for the Future Inc.

Simmons, I.G. 1966: 'Wilderness in the mid-twentieth century U.S.A.', *Town Planning Review* **36**, 249–56.

Triggs, G.D. (ed) 1987: *The Antarctic treaty regime. Law, environment and resources.* CUP Studies in Polar Research.

Vicuña, F.O. 1988: *Antarctic mineral exploitation. The Emerging Legal Framework.* CUP Studies in Polar Research.

Wolf, E.C. 1985: 'Conserving biological diversity', in L.R. Brown (ed) *State of the world 1985.* Washington DC: WRI, 124–46.

Wolf, E.C. 1988: 'Avoiding a mass extinction of species', in L.R. Brown (ed) *State of the world 1988.* Washington DC: WRI, 101–17.

Wastes

Battarbee, R.W. *et al* 1988: *Lake acidification in the United Kingdom 1800–1986.* London: Ensis Publishing.

Bedding, J. 1989: 'Money down the drains', *New Scientist* 122, 34–38.

Clark, R.B. 1989: *Marine pollution.* Oxford: Clarendon Press, 2nd edn.

El-Hinnawi E. and Hashmi, M.H. 1987: *The state of the environment.* London: Butterworths for UNEP.

Elsom, D. 1987: *Atmospheric pollution.* Oxford: Basil Blackwell.

Flavin, C. 1990: 'Slowing global warming', in L.R. Brown (ed) *State of the world 1990.* Washington DC: Worldwatch Institute, 17–38.

French, H.F. 1990: *Clearing the air: a global agenda.* Washington DC: *Worldwatch Paper* 94.

Gribbin, J. 1990: 'Warmer seas increase greenhouse effect', *New Scientist* 125, 31.

Holdgate, M.W. 1979: *A perspective of environmental pollution.* CUP.

Houghton, J.T., Jenkins, G.J. and Ephraums, J.J. (eds) 1990. *Climate change. The IPCC scientific assessment.* Cambridge University Press.

Kemp, D.D. 1990: *Global environmental issues. A climatological approach.* London: Routledge.

Leggett, J. (ed) 1990: *Global warming: the Greenpeace report.* OUP.

McCormick, J. 1989: *Acid earth. The global threat of acid pollution.* London: Earthscan Publications.

MacKenzie, D. 1989: 'If you can't treat it, ship it' *New Scientist* 122, 24–25.

Park, C.C. 1987: *Acid rain: rhetoric and reality.* London: Routledge.

Rhode, B. (ed) 1988: *Air pollution in Europe.* Vol 2. *Socialist countries.* Vienna: ECCRDSS Occasional Paper no 5.

Part III

GLOBAL OVERVIEWS

9

The regionalisation of interactions

An unequal world

A look at any atlas shows an unequal distribution of everything. Even the simplest map reveals particular patterns of high and low ground, or different sizes of nation states. Interpretive atlases with maps of income or nutritional status emphasise even further these inequalities; we may even be implicitly invited to think that they are all as inevitable as the location of mountain chains or explicitly told to work for change towards a greater evenness of shares in all material things.

That individual nation states occupy different endowments of land surface, shore and ocean zones, and climatic inputs may seem so trite a statement as not to be worth its space. Yet we may need reminding that the physical resources of a nation are still significant in its inhabitants' life-style even in an age of trade, development aid and instant telecommunications. It is a gross display of environmental determinism to suggest that it is due to climate that the developed nations of the world lie in the north temperate zone, for example, but less so to point to the combination of coal and metallurgical development that allowed their early development into a position of power which they have been able to maintain. So climate, soils and landforms, precipitation and temperature are still major elements to be assessed in any nation's attempts to maximise its material assets; even the richest cannot escape the boundaries of a long cold winter or the occurrence of earthquakes, for example. In one direction, this uneven heritage leads globally to a dichotomy of opinion as to whether every nation should direct its agricultural production to raising those crops which flourish best there and exporting them to acquire currency to buy in other necessary foods and fibres, or whether some degree (even 100%) of self-sufficiency within the borders of the nation or trading bloc is essential for security.

When we look at the outcome of human-directed systems then uneven standards are rife: any map of almost any measurement reveals large gaps between those who enjoy the highest levels (of health, income, nutrition, military protection, access to wilderness or whatever) and those who come at the bottom of one or more (or most) league tables. None of these sets of data measure anything beyond material phenomena, of course, but only the most unreconstructed romantic would argue that high levels of disease and malnutrition, poor housing and education, or the powerlessness of most individuals, are somehow a desirable and dignified state. The people may be dignified indeed but in spite of their material conditions, not the other way round.

Uneven levels of social and economic development are paralleled in political terms. Industrialisation has meant economic and military power and so it is not the small states of Andean America or the South Pacific who determine the actual course of global affairs in those times when large-scale decisions are made, as in recent responses to threats of global climatic warming. Thus such discussions tend to take place outside the framework of the UN since the small and the less developed have the majority vote in that body. Only when the poor and the small get together in a bloc, as over the future of Antarctica, can they exert much influence: the example of OPEC in the 1960s and 1970s has been strong. Most such countries must dream about the benefits of being the sole repository of a resource that every rich nation must have, (rather as the Duchy of Grand Fenwick, in one of Peter Sellers' better movies, *The Mouse that Roared*, possessed the only example of a super-bomb and in a roundabout way blackmailed the US into filling their coffers) such as a plant product that would catalyse nuclear fusion reactions in small quantities. But in the fields of resources and environment as in many others it is the large and powerful who set the pace for trade and exert a hegemony over ideas (about 'development' and environmental management for instance) which is reinforced by the technology of communications which they are able to emplace.

Reactions by the powerful to this uneven situation are, grossly, two-fold. One set suggests that nothing that industrialised nations can do will be any help in the long run and that disadvantaged countries are best left to themselves to sink or swim. The other, stronger, set believes in the existence of a moral responsibility (fuelled in part by a sense of guilt about colonialism, in part by a fear of the appeal of 'Communism' or other fundamentalist movements among the poor, and in part by simple pity) which finds its way into the whole process of 'development'.

Developed and less developed

Many books rely on a simple division of the countries of the world into two categories: the developed (DCs) and the less developed (LDCs) and this is useful enough so long as precise definitions are not necessary and a broad category of conditions is being referred to. More complicated official categories of rich and poor are however widely used and have relevance for resource use and consequent environmental changes.

The most generally adopted classification uses income (measured as GNP per head in US dollars) as the defining criterion, with the major groups being:

- Low-income economies (total population 1987, 2,822 million) in the income range $130–450, average $290. Ethiopia is at the lowest end of the range and Liberia at the top (Plate 3.1); Somalia, Togo and China represent the mean. India ($300) is also in this group.

- Lower-middle income economies (total population 1987, 610 million) in the income range $520–1930, average $1200. Senegal is at the foot and Poland at the top, with Turkey at $1210 being nearest to the average. South Africa ($1890) falls in this group.

- Upper-middle income economies (total population 1987, 435.5 million) in the range $2020–5810, average $2710. The lowest figure is represented by Brazil and the highest by Oman, with Gabon at $2710 being very close to average. Portugal ($2830) and Greece ($4020) belong to this set.

- High-income economies (total population 1987, 777 million) in the range $6010 (Spain) to $21,330 (Switzerland), average $14,430 (West Germany was $14,440). The UK is $10,420; USA $18,530 and Australia $11,100.

Plate 3.1 The low-income economy in Mali, 1985. Preparation of millet by hand. No immediate supplies of power or water and little evidence of material possessions. Something to leave behind as soon as possible or an invaluable ethical and behavioural repository?

These data say nothing, of course, about the actual distribution of wealth in the nation concerned: there may be pockets of high-income living in a largely undeveloped country as happens in some of the oil producers of the Middle East, whereas in socialist economies the gaps between high and low incomes are much less.

A few general conclusions can be drawn from these data. The gap between rich and poor is very wide indeed and that the ongoing pattern is that the poor are gaining a large proportion of the population growth and the rich ever more of the income. The relation to resource use is more complex but if we use energy as a surrogate for resource use then a linear relationship between GDP and energy use is apparent. This however only plots commercial energy, so that the situation of the woodfuel users is somewhat exaggerated. The rich countries use their wealth to get resources on a world-wide basis; the poor are restricted to local sources so that their search for the material basis of living must be very intensive. But they may also be the recipients of the HIEs' outreach for materials; thus the LIEs and LMIEs are likely to be the places in which the main environmental impacts of resource production (though not necessarily that of consumption) are felt (Plate 3.2).

Brief mention may also be made of spatial distributions. Overall, the richest countries are in the temperate belt of the Northern hemisphere, a result of both endowments and cultural history. East Asia is home to a very large number (1513 million) of low and middle (both groups) income people, as is South Asia (1081 million). Of the 20 nations below $300 GNP/cap, 15 are in Africa south of the Sahara and the rest (which include China and

Plate 3.2 The high-income economy: Woking (South England) from the air showing large houses at a low density, raising the costs of providing them all with gas, electricity, sewage and water and rendering inevitable the extensive use of private autos. Unless walking raises the sort of sweat that needs a big pool to wash it off.

Bangladesh) are in Asia. Only Japan, Singapore and Hong Kong from Asia join the HIEs, in which group there are no countries from Africa or Latin America. Hence the economies outside the HIE group are sometimes called the South.

Towards development

What have these patterns meant in the recent past for resource use and environmental change? The gap between rich and poor has been of concern to many official and philanthropic groups and the usual response in the last 70 years has been that 'development' towards North American or European ways of gaining livelihoods (and by extension, patterns of resource use) represents the 'solution' for their precarious positions. Since industrialisation generally represents the departure point of the growth of HI and L/MI economies in the 19th century, the obvious conclusion to draw, especially in the period of de-colonisation after 1945, was that the installation of industry on the Northern (or Western) pattern was the answer. It had, after all, transformed Japan in the short time after the Meiji Restoration of 1868 into a major military power capable of virtually annihilating the Russian fleet in 1905 and then in 1941 doing much of the same to the USA at Pearl Harbor. So a sequence of stages (in which the transfer of western technology is always important) was deemed to be necessary: first the build-up of income from exploiting a large-scale mineral deposit or plantation crop, then the manufacture of goods in order to avoid importing them and thereafter the construction of heavy industry, centred upon steel. The fuels would preferably come from indigenous hydropower or coal or from cheap oil out of the Middle East. Eventually there would be an export-oriented economy from the industrial base. This type of growth would absorb population growth and stimulate agricultural progress by creating large urban markets.

The appeal of such linear sequences was such that they became the staples in development theory. Organisations such as the World Bank used them in directing capital and organising projects within the receiving countries: large dams for hydropower to power industry were especially popular since *inter alia* they provided a lot of employment for unskilled workers. The fact is, though, that this type of orderly transition has (a) not always worked and (b) where it may have done so, there have been very undesirable side-effects. Renewable resources have been exploited very rapidly, non-renewables brought to the point of depletion, and environmental contamination taken to very high levels since all were held to be worth paying such a price for development. Social disruption has also followed since the cities became a magnet for poor rural inhabitants who were being dispossessed by those able to use capital to modernise food and fibre production (often for export) in ways that needed less labour. A further drawback with such a model is the energy required: add up all the plans for industrialisation in the South during the 1970s and the fossil fuel energy requirement far exceeded anything that the earth's crust could supply. Where indigenous energy was not available, then a nation's development was hooked to the price of oil on the world market, itself a product of politics rather more than economics.

Alternatives (especially those applicable in the tropics) have been slow to emerge and be adopted, since large organisations like the World Bank are prone to institutional inertia, and the companies which actually control affairs on the ground are not keen to radicalise away their profits. Development alternatives have centred around the following kinds of strategy:

- The primacy of local and regional self-sufficiency for basic needs rather the dominance of the tie into world trade patterns whose terms are controlled by the Western countries.

- The primacy of environmental considerations: all development to be as little disruptive to the biophysical environment as possible, with the use of small-scale 'appropriate' technology wherever possible, especially in the use of energy.

- The re-evaluation of indigenous knowledge, recognising that the residues of pre-industrial lifeways may provide the seedbed for economies less geared to satisfying externally-generated demands.

- The central importance of the economy of the rural areas, both as producers of necessities and retainers of population, lessening any impact on the cities. Thus high-technology 'Green Revolution' type developments are viewed with suspicion.

- The need to give power back to the inhabitants of the land and away from the dominance of MNCs, landlords both present and absentee, and government agencies obsessed with implanting western ideas and technology.

- The recognition of the central importance of women, especially in rural areas, and the reflection of this in their standing in the community's decision-making procedures.

The short-hand term for this type of development is 'bottom-up' change and it is gaining in strength, especially where local groups have decided to make their own futures. The scope for conflict at all levels is high and change slow but seems to be in tune with many currents in the world. Two key themes addressed in an integrated way by small-scale development are the role of population growth in the process of bringing about a better and more dignified life for poorer people, and the notion of development that looks beyond the next year's food supply or balance sheet, i.e. the idea of sustainability; this latter is dealt with below.

During the hegemony of 'top-down' development thinking, population growth was regarded as a principal cause of the slowing-down of progress. If growth of population was fast, the capital accumulation per capita was lower and so the take-off stages into full industrial status would be delayed. Since industrialisation in the West had been accompanied first by a surge in growth rates and then a rapid fall, it seemed clear that the proper course was to trim the rapid growth rates of the LDCs by making available all the Western technology of population control to the lower income countries, with modifications to the method mix according to local susceptibilities and customs. The availability during the 1960s of oral contraception was seen as presaging a breakthrough: parachute the pill into every rural community and 1% per annum increase rates would soon be the norm in South as well as North.

There is no denying that these family planning campaigns have had some effect, not least in the sense that the world rate of population increase has fallen from a 1960s high of about 2% per annum to a present level nearer 1.7%. Part of the fall in world rate is of course due to continuing drops in growth rates in the DCs, some of which are now down to zero growth or even negative rates. Some low and middle income economies have never-

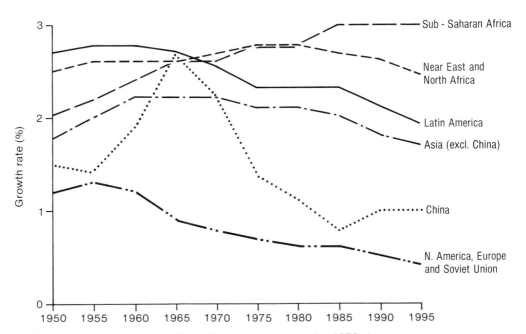

Fig. 3.1 Regional trends in world population growth since the 1950s in per cent per annum. Source: US Department of Commerce census Bureau, *World population profile 1985*. US Govt Ptg Office, Washington, DC.

theless cut 1–2% off their growth rates, though there are still many with 3 or even 4%, which equal doubling times of 23 and 17.5 years respectively (Fig 3.1). The achievement of smaller family sizes is seen in most government circles as a desirable objective, though not necessarily the only development goal worth pursuing. There are still some aggressively pro-natal governments in the world, and a central dogma of the Roman Catholic church is to oppose any methods of limiting conception that depend upon chemistry and physics rather than mathematics.

But the reversal of the thinking of the 1960s is perhaps encapsulated by the famous remark of the Chinese delegate at the UN Population Conference in 1974 that, 'development is the best contraceptive'. That is to say, that smaller families follow the achievement of decent levels of nutrition, health, shelter and education rather than being a pre-condition for getting them in the first place. While people are still insecure about their future then children are the only investment whose value can be relied upon. Security beyond the immediate future however allows individuals to make other choices, though these may still include large families either by design or in the cause of making sure that one child is male, still a desire frustrating even the thorough-going family-limitation measures adopted in China.

It has been relatively common in the 1980s, therefore, to demote population control from a dominant position in the development debate to a more equal position along with other problems. Nevertheless, it cannot be ignored since the number of mouths to be fed and hands to be employed are still the driving force behind much of the resource use and the environmental change in the world. And since the rich use more resources per head than the poor, the latter have very high expectations of a materially good life in which they will have a share.

Sustainable development

The general picture of resources in development is that of using them as fast as feasible either to make profits or to stay alive. Yet it is clear to most observers that such short-term tactics will lead to collapse of profits or of productive ecosystems or both; yet the large company has the option of getting out before too late whereas the peasant does not. Hence the search for sustainability in economic development. This has been largely motivated by those concerned on behalf of indigenous people but is now also a matter for consideration by large corporations and sources of finance.

Sustainability has come to be a symbol as well as a word, denoting a whole syndrome of approach to economic development which emphasises the ecological rather than the economic. It is as well, therefore, to remind ourselves of how its meaning is derived and applied, since it appears that both ecology and economics have contributed to its meanings:

- The science of ecology has put forward the notion of resilience, which is the ability to maintain a coherent structure and function in the face of external pressures ('stress') for change. In the case of agro-ecosystems we mainly think in terms of the system's capability to maintain output in the face of lowered mineral nutrition levels, erosion, toxic chemicals or salinification, for instance. Sustainability then addresses itself to the process of maintaining the system so as to reduce stresses and maintain a stable output; in essence a form of equilibrium.

- The discipline of economics has given us two sub-concepts in this field:

    Sustainable economic growth, which is the growth of per capita incomes maintained over time; and

Sustainable use of resources and environment, which implies some rate of use of the environment and its resources which does not deplete its capital value.

Put together, the essential features of economic notions of sustainability seem to be living off the income of any process rather than the capital, i.e. not depleting the wealth-creating ingredients like resources stock, technology, and knowledge. In chronological terms, the processes must also conserve intergenerational equity (i.e. options must not be closed off to future generations) and offer resilience to stress. In an ideal world, equity between people now would be a desirable element as well, especially in the poorer countries.

What meaning is there in this concept for the development of the LIEs and the LMIEs? It looks as if they have two hypothetical choices:

- A very large-scale relocation of populations combined with huge donations of capital and income to relieve pressure on the environment by buying in food and energy.

- Attention to regeneration of the flow resources of the environment as a preliminary to finding out how to manage them so as not to deplete their value as capital. 'Finding out' may be a political process just as much as a scientific one and in any case it is not a cheap alternative compared with the aid-intensive pathway mentioned above.

Neither of these is easy, neither is inexpensive. The first ties nations to external values and materials but the second threatens revolution to all kinds of vested interests. It also implies a pause in economic growth while the ground is prepared for sustainable resource and environmental use, with all the difficulties thereby implied.

It is clear that the conditions for sustainable use of resources and development are rarely fulfilled in the LIEs and LMIEs and probably not further up the income scale, either. What might help in the approach to even the first stage of regeneration of the ecology to provide a base for sustained use? A number of possibilities, none of them exclusive, are usually discussed:

- The use of more technology imported from North to South. The problem with such an agent of change is that advanced technologies usually contribute to social inequality; the 'Green Revolution' is normally held to be, overall, an example of this process. Equally, the Brazilian programme to produce alcohol as a fuel from sugar and cassava plants has led to lower rates of output of staple crops and domestic foods and to regional inequalities. The next 'Green Revolution' is widely held to be that of biotechnology; if this is mostly genetic engineering then it seems likely that the main benefits will go to the source areas of the technology, who will after all hold the patents on the strains of plants and animals involved. Will any diminished dependence on the 'natural' environment necessarily lead to benefits for the impoverished?

- The reform of an economics in which neither the neo-classical capitalist nor the Marxist positions offers immediate solutions to the negotiation of a resource use pattern which has the promise of sustainability, though clearly the discussion above shows that there is still flexibility in both. The main practical step seems to be the re-alignment of the terms of trade between rich and poor in which inevitably the poor are always disadvantaged. Numerous reports, including the high-profile Brandt Commission (1983) have argued this but without producing any major shifts as yet.

- Reform of economics can only stem from political change. Questions of scale arise: internally the action of citizen groups in attempting to wrest power from indigenous but exploitive masters, or agents of MNCs, is growing but must still be recognised as being localised in nature. National revolutions at present rarely seem to clear enough space for different choices about resource and environment since survival is still as much as can be offered. In the international sphere the radical position is that the HIEs should de-industrialise, since they over-consume and over-produce. If they were to scale these activities down, then the reduction in the demand for materials and energy would enable more of these to be devoted to meeting basic needs in the poorer places. The only trend in this direction might be part of the response to global climatic warming, i.e. as a response to threatened self-interest rather than an act of philanthropy.

These discussions are reformist but not radical. Beyond them, more thoroughgoing principles for development are canvassed. These are based on notions such as social and cultural integrity of development, i.e. the growth of development from within. To this is added solidarity and emancipation, and non-violence in both its direct and structural senses. All these must be set in an error-friendly context; perhaps this is the time to be reminded that there is still a basic envelope which must provide any sort of sustainability. This is composed of NPP, which reflects the fixation of renewable solar energy and such materials as can be sold, recycled or exchanged for knowledge of how to do without them. These limits, at all scales, have not yet been transcended by any political or economic voodoo although the popular perception (upon which many a stock market rests) may tell us very loudly not to worry.

10

Management frameworks

How does it happen?

A good deal of resource use and environmental management in the world is done by individuals or small groups of people, often acting in accordance with customs which are orally transmitted. But beyond them, there is a great body of written policy and law which affects resource processes. It is all very well for us to lay out a series of criteria, as in I.3, which differentiate the ecology, economics and behavioural characteristics of resources systems but in real life, action is taken in a fashion which reflects a holistic outcome of all of these (and maybe others as well, like ignorance) laid out in the form of written statements about policies, laws and institutions.

World-view

Though not explicitly stated, the existence of certain principles of behaviour set in a very wide framework are important in determining patterns of resource use. Dominant among these is the collection of themes that comprise the western worldview; the core idea is that the earth is a set of materials for human use, that material conditions are expected to get ever better for people, and that technology is the key to providing that affluence. In general, more of most materials (and certainly more choices), constitutes progress. If we want a simple label, then perhaps *technocentric* would be a reasonable word. This world-view is common to both free-market (capitalist) and socialist economies. Further, with the spread of Western imperialism in the 19th century, reinforced by the economic and cultural dominance of the West in the 20th century and now consolidated by the primacy of the industrial nations' media moguls in the field of global communications such as satellite TV, these western views hold sway over much of the earth. They influence (and will do so even more in the next 20–40 years at least) the outlooks and attitudes of people even in remote places and living in relatively simple economies. Hence although it may have been possible in the past to maintain economies anchored in other points of view, like the sacredness of nature or the fulfilment of only a limited range of material demands, it seems unlikely that these will persist. The western world-view is something of an umbrella for all that follows in this chapter.

Since the ideas of the West are so over-arching, it becomes important that the dominant nations evolve and transmit alternatives to the current technocentric world-view, and the last chapter of the book will be given over to looking at some of the alternatives which are the currency of those in the West with radically different viewpoints. However, these have not in general found their way into the codes of behaviour which are laid down by most institutions that deal with resources and environment, and with the nature and effectiveness of which we must now deal.

National governments

In the web of organisations which the modern world has erected to deal with its resource processes, it is not simple to find a logical starting-point. But since the nation state is such a fundamental building-block of most legal structures, we will begin there and work as it were outwards.

The basic process is the adoption of a set of policies for resources and environment which are then transformed into law. This is then administered through manifold regulations which derive their character and validity from the written law: i.e. they can be amplifications of the law but must not be in conflict with it. One great difference between different kinds of government is the degree to which they attempt to influence resource processes and consequent environmental change. Those with a commitment to collective modes of action, such as socialist governments, generally wish to regulate most phases of the systems, by having laws about emissions for example, or by having major resource processes in nationalised control, as with energy, manufacturing and agriculture. Free-market capitalist governments, on the other hand, believe in a minimal role for government and so resource supply is usually left to the dictates of the market. Ideally, a properly costed system would also deal with environmental change. Environmental taxes on e.g. carbon and nitrogen emissions are often proposed, for example. If the UK put a tax of 15% on domestic fuel then it might reduce demand for energy by 5½%. But then the poorest tenth of people would pay £1 more per week in this tax and have to cut consumption by almost 10%, whereas the richest tenth would pay £2 more in tax but being richer anyway would cut consumption by less than 2%. A 40% tax on nitrogen fertilisers would mean a cut of 27% in farm profits but the concentration of nitrates in groundwater would only be reduced by 5%. In the absence of any agreed pricing of external costs, it is usually necessary to legislate for emission levels and the control of toxic wastes, for example. (This is not to say that economists are not tackling the problem of attaching prices to environmental features and processes.) In the case of radioactive wastes, no government can abdicate its overall responsibility since the danger to life of some of these substances is likely to outlast that of any private company, whereas we assume that the nation state will last as long as our species.

Government policies can then be divided roughly into two groups:

- Those which are *for* things: e.g. an energy policy which encourages the use of one source rather than another, or which supplies it at particular levels by means of rationing, as in Romania in the 1980s where power and heating was supplied for only so many hours per day. Food policies might do the same kind of thing and could encourage the eating of less meat in the West or by contrast could be oriented by means of subsidies to preserving e.g. sheep rearing in marginal environments. Similarly governments can do much to influence family size by pursuing pro- or anti-natalist policies.

- Those which are *against* particular kinds of processes or events. Environmental pollution is the obvious example, and most governments have some form of regulations (not always obeyed) in this field. Industrial development location is another matter which may not always be left to the free market: there may be incentives to locate industry (and with it the complex links of resource supply and environmental metamorphosis which normally accompany it) in areas of high unemployment for instance or as a way of claiming territory.

Few nation states, however, are isolated islands and so they have to manage their territorial resources and environments in accord (or not) with neighbours. Many environmental problems cross national boundaries and so regulatory levels above those of the state are often needed.

Supra-national organisations

Nation states come together for many purposes but for our theme there are two main areas for consideration. The first of these is the conclusion of one-off agreements for a specific purpose like managing an internationally-distributed resource, and the second is a general-purpose supranational body like the European Community which has to deal with environment and resources just as it must act on foreign policy, trade or transport. In addition, a block of nations may be the subject of special concern by a sub-division of a higher unit, as with the Regional Economic Commissions (e.g. for Europe, Africa *et al*) of the United Nations.

The first type has a long history in the sense of managing common resources. Even in the 19th century, Pacific countries concluded agreements to prevent the extinction of fur seals, and there has been a subsequent record of catch limits on certain kinds of fish; these bi-lateral or multi-lateral accords were especially necessary before the designation of 200-mile EEZ's (see p 112) when the oceans might be exploited by anybody with suitable technology. When a commons existed, it was vital for the main users to regulate themselves so that newcomers could be policed or excluded altogether. The same has been true of anti-pollution agreements, as with the attempts of Canada and the USA to clean up the Great Lakes; neither one could achieve this alone and a coordinated approach has proved to be essential. Analogous processes underlie discussions about the control of acid deposition in the south-east/north-east of these two nations but agreement is proving hard to achieve.

Multi-purpose organisations can have a great effect on resources and environment, both by means of direct policies and indirectly as the results of other actions. Thus the EC, for example, may have direct effects on drinking water standards through community-wide pollution control laws but may have even greater indirect effects via agricultural policies which encourage intensive agriculture and hence the application of large quantities of nitrogenous fertiliser to the land. Environmental change is likely to be speeded up as the result of regional policies which bring about investment (Plate 3.3) in areas otherwise relatively poor: the industrialisation of parts of northern Portugal, with effects on land use, water quality, and shoreline management may be quoted here. Such organisations are really much like a national government but on a larger scale: they can deal with trans-boundary problems better then concordats between several states in the sense that frameworks already exist and do not have to be specially set up. It seems likely that only the EC could bring harmony and consistency into the highly variable mix of 'green taxes' proposed by European countries. Some want none (France, UK), others very few (Italy has a special 100 lire tax on plastic bags) and yet others intend to be thorough, like Denmark and the

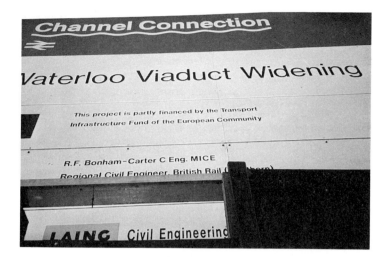

Plate 3.3 The role of economic groupings in resource development and environmental change is shown in this billboard in London. Fast train connections between London and Paris via a tunnel will increase the pressure for economic development in selected areas of both countries.

Netherlands with proposals for carbon-emission imposts. On the other hand, bodies like the EC may be slower to respond to new or newly-discovered environmental problems than an individual state, not least because they have to be slow to contravene national sovereignty except where such rights have been explicitly given up.

So as with nations, policies and laws may be both for things and against them. Pollution is the most obvious target for the direct exercise of the 'against' variety: standards can be set and then enforced. Air and water quality are good examples of early targets in such a context. In terms of being for the supply of certain resources, supranational organisations are usually more cautious but the EC, for example, has quotas for agricultural production which aim to eliminate over-production of certain foodstuffs and an agreement to share energy supplies if there is a sudden shortage resulting from e.g. politcal decisions in the Middle East. The outstanding example in all these fields is the EC but other blocks of nations are coming to see the advantages that can accrue in resource and environmental matters.

Global organisations

Many large companies are in fact global organisations but here we shall deal with public structures, of which the main example is the United Nations Organisation (UN). Also important are international finance bodies such as the International Bank for Reconstruction and Development (IBRD) usually called the World Bank. We can classify the activities of these world-wide phenomena into roughly two groups: (a) those that get involved on the ground showing how things can be done, and (b) those that store data and transfer information or money and thus act as catalysts but do not themselves get wet and dirty.

The first group is typified by some of the agencies of the UN which are concerned with development. They maintain databases and produce innovative research and publications but they are also concerned with sending out teams of experts who advise on the ground as well as in government offices. An outstanding example is the UN Food and Agriculture Organisation (FAO) which focuses its concern on the improvement of nutritional standards. It maintains a central HQ and regional offices, gathers information and

Plate 3.4 All the techniques of advertising are used (in 1988) by environmentalist groups in campaigns to change attitudes. Directing consumption is one of the immediate targets, a lesson learned from more overtly political boycotts.

conducts research, and puts together teams of appropriate experts to improve e.g. soil erosion problems or protein deficiencies, or livestock genetic traits. The UN World Health Organisation (WHO) is analogous in its field and is closely allied in structure and methods to the UN Fund for Population Activities (UNFPA) which is concerned not only with spacing births but with maternal health and sex education, for example. The tasks of all these groups is very difficult, not least because they can be seen to represent 'top-down' development in which Western/Northern technology (be it pesticide or pill) is applied to the South as the answer to the problems as perceived by the agencies themselves.

Behind these on-the-ground agencies lie those who set a broader context by trying to get global pictures and wide overviews before advising UN agencies or national governments. Set up in 1972, the UN Environmental Programme (UNEP) is of this type. Its main function is to coordinate the efforts of all UN agencies with an environmental concern, to extend the range of data available on environmental matters and to highlight areas of problem or potential problem. It operates through established channels and does not have local teams. Regrettably, its budget has been reduced over the years from $100 million to $30 million and is thus greatly hampered in its task.

More directly involved at one level is the IBRD, which provides capital for development in the South. Many large projects, especially those concerned with water control and electricity generation, have been funded by the World Bank. It has thereby been heavily criticised for its ignorance of the environmental and social consequences of some of its projects. One response in recent years has been to question the financing of large dams, for example, and to look with greater sympathy upon needs generated by, and expressed at, grass-roots level. Indeed, some of its major projects (Palonoresti in Brazil, Three Gorges Dam in China, cattle ranching in Botswana, transmigration in Indonesia and irrigation in the Narmada valley of India) have been dubbed 'the fatal five'. After 1987, the Bank expanded the role of environmental impact assessment in its projects, though there are doubters about the seriousness of its intent. But overall, IBRD has been an instrument of technology transfer from North to South. In this respect its critics regard it as an agent of Western production and profit, opening up the Third World to international private enterprise. Similar views of the International Monetary Fund (IMF) have been expressed: many desirable subsidies, e.g. for education and food have had to be abandoned in LIEs in order to promote the export of, for example, tobacco. Trade, it may be noted, may militate against environmental protection. GATT (General Agreement on Tariffs and Trade) contains limitations on governments' powers to use subsidies or erect tariffs which might, as one result, be applied to environmental care. GATT has in particular facilitated the use of LIEs as dumping grounds for toxic wastes. The wording of GATT itself does not include 'environment'.

Overlapping the concerns of all of these has been the International Union for the Conservation of Nature and Natural Resources (IUCN) which initially was largely devoted to the international aspects of species conservation but which has become the leading protagonist for sustainable development based on renewable resources, without distinction for North or South. IUCN has also fostered the idea of debt-for-nature swaps. These involve the purchase of a developing country's debt and its cancellation in return for environmentally positive action on the part of the LDC. The Worldwide Fund for Wildlife (WWF) has itself engaged in such purchases in Ecuador and Costa Rica and other nations are interested, especially in Latin America.

Local government

In many cases, the local government units of a national system see the sharp end of resource and environment issues. They are often in the front line, so to speak, of any conflict between the citizens who are to experience (for good or ill) any changes that occur. Yet they may not have initiated these policies and indeed be actively opposed to them (as when for example a nationalised energy industry proposes to open a new strip mine for coal in an agricultural area) and so are the unwilling deliverers of policies conceived without their consent. Local government is a complex entity, varying enormously in its structure and responsibilities from nation to nation but it often carries out two particular tasks germane to our theme:

- *Development control* This is usually part of the physical planning process now common in DCs. Here the local government unit, acting within the context of national or supranational legislation, decides on matters such as land use zoning: which areas are to be devoted to productive uses like industry, housing and commercial development, and which are to be in a protective use such as watershed area, parks or nature reserves. This may not all be as negative as it sounds since a large measure of economic planning may occur whose function it is to develop particular resources. In DCs these most likely will be social, e.g. to provide employment but may also involve environmental considerations such as siting small factories in attractive surroundings.

- *Waste management* In general, national governments do not wish to get involved in the day-to-day aspects of these processes except where the most dangerous materials are concerned, such as radioactivity and discarded munitions. At the same time, the private sector has in places been judged incapable of sustaining a commitment to the proper environmental considerations needed for the safe treatment and disposal of 'normal' wastes. which include some very toxic by-products of industrial processes. Local authorities, then, have to deal with municipal wastes. sometimes sewage, and some industrial wastes. Protection of local waters, the finding of local burial sites, and local air quality are likely thus to be the responsibilities of the town, county or regional council. As such, they are constantly being badgered to improve their standards by citizens and citizen groups which want higher environmental quality. On the other hand, major taxpayers such as industry and commerce, or sources of subsidy like a national government, want less money spent on these matters so as to keep down corporate or public expenditure. Smaller local governments, too, are not likely to attract either the elected representatives or the professional staff to deal well with such complex interactions of public perception, scientific information and financial constraints.

The private sector

So far we have looked at the role of various organisations of a public character. Much of the actual processing of resources is however carried out by firms in the private sector along with their analogues in the Centrally Planned Economies. So companies are the resource finders, developers and suppliers. Often they operate on a very large scale across several nations, if not continents and so the label Transnational or Multinational Company (MNC) is applied. Examples can be found in the areas of metals, timber, rubber, food and energy as well as many others, and international waste disposal is now entering the list of flows thus controlled.

The great virtues of MNCs lie in their capacity to absorb rapid change. In the energy field, for example an oil company with developments in many different places can ensure a constant flow to its customers. If one source dries up temporarily because of a revolution or because a refinery is out of action, then there will be surplus capacity elsewhere which can step in to keep up supplies. If the coffee harvest in South America is blighted by frost then Megafoods Inc. can supply its Dutch outlets with coffee from Tanzania without any hiccup in the flow other than an increase, no doubt, in the price to the consumer.

Because of their size, the MNCs can also cause difficulties for the nations in which they operate. A mine for instance in a small LDC may provide a great proportion of the local employment and GDP but for a MNC it is a small operation, to be closed if the price for the metal falls slightly, or the costs at the mine get higher due to environmental protection measures or higher wages. Equally, MNCs may be able to manipulate demand, as was shown classically by J.K. Galbraith. Advertising especially can be used to create and maintain a demand for a product (like powdered milk in LDCs or frequent changes of car in DCs) which back along the chain of resource process and must increase environmental impact.

The same can be true of many firms which are not multinational. They too can manipulate demand and they too are responsible on the ground for resource supply and environmental change. They can also be the technological innovators, more willing to take risks than big bureaucracies, who may for example, pioneer new processes which lead to lower environmental impact or to greater energy efficiency in a process. It is a mistake to see all private corporate involvement with resources and environment as being detrimental to everything except the company profits, just as only the ideologically blind would deny that all levels of government have a key role as arbiters of the common good.

The commons

In 1968, the biologist Garret Hardin pointed out that common-property resources such as the open sea and common grazing land were subject to most abuse. Everybody gained from extra use by a small amount but inevitably there came a crash when the resource could no longer bear the level of exploitation to which it was subjected. Cures for these conditions seem to be two-fold. The first is passage of the resource into private owner-ship as has happened with the 200-mile EEZs in the oceans. It is difficult to see how the atmosphere might be so parcelled, however, to quote a current example of concern. The other is to remember that self-regulation is also practised by some users of common resources: wildlife hunters, lobster fishermen, irrigation users and communal forest dwellers are all cited. These latter groups are usually small and 'traditional' in life-style but give us hope perhaps that larger units might learn that cooperation is as feasible as competition.

NGOs

Ranged around the twin columns of the modern world economy, those of state and business, are the organisations founded by private citizens to pursue other ends; they pride themselves on freedom of thought and action and see themselves as collective consciences for those organisations (of any type) who seem to lack such a facility, and generally as goads of all those who are deaf to all calls except those of profit or ideology. They are by no means confined to the environmental field (development has a particularly dense representation) but it is an area which particularly spawns them. Like other organisations, they can be small and oriented towards a single-issue or be virtually multinational com-panies. We can distinguish two main types:

- Those that are involved upon the ground with managing a particular resource or ecosystem. In the UK, examples are County Conservation Trusts which have nature reserves, and national organisations like the RSPB which also manage land and water resources for the perpetuation of wildlife. These bodies are usually involved in pleasure and protection rather than production, though angling clubs may take home some of the catch. All are willing to engage in campaigning when their interests are threatened by actual or proposed environmental change.

- Those that do not own or lease land themselves but are campaigning bodies who harry governments or industries and commerce for better environmental practices. The Worldwide Fund for Wildlife, Greenpeace and Friends of the Earth are examples. Their membership has risen greatly in recent years, as shown for the UK in Fig 3.2. Such groups have usually built up great stores of expertise so that they can challenge the accepted view on any particular topic: for example on the issue of civil nuclear power they can produce their own economists to dispute the costings of the industry itself; on whaling they can argue that 'scientific' whaling is unnecessary for population studies by producing biologists who affirm the existence of non-destructive methods of age-sex determination of minke whales. Some, like Greenpeace, apply the non-violent direct action techniques of civil disobedience movements through the years.

Those who envy these groups' independence, but who feel rooted in a more establishment-minded tradition of change tend to set up permanent or *ad hoc* bodies of

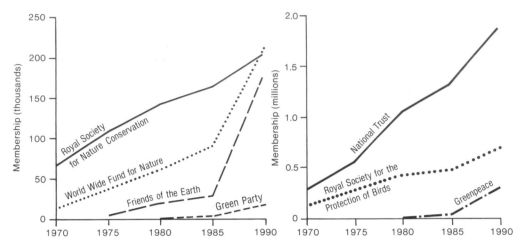

Fig. 3.2   The pace of growth in the Green movement in the UK as shown by the membership of a number of organisations during the period 1970–90. Source: *The Economist* March 3, 1990.

the great and the good to produce reports which catalyse transformations of attitudes and practice. The International Council of Scientific Unions (ICSU) for example, has set up SCOPE (Scientific Committee on Problems of the Environment) which has produced a series of volumes of expert studies: those on global biogeochemistry and 'nuclear winter' have been especially influential. Following the Brandt Commission's look at the economic context of North–South relationships, the Brundtland Commission tried to do the same for the environmental aspects (see p 217).

The individual

It is a commonplace of today's world that the individual feels powerless. Complex governmental structures remove people from the sites of key decisions and even where they have representatives, such individuals may themselves owe their position to a party whose demands are more powerful than those of the electors. Equally, the idea of the sovereignty of the consumer is arrant nonsense when many of their demands in the DCs are orchestrated by large companies. No 'consumer greenery' is going to solve main problems since the time environmental relations of consumer products (including the energy costs) are rarely known at the point of sale. The recovery of individuality is a complex matter and not one to be solved by nostrums in textbooks. But the ways in which we define ourselves are somewhere at the heart of the issue of the relations of humanity and nature. At the end of this book they will have to be brought out again and examined.

Conclusions

We are dealing with a complicated picture, with many overlapping layers as well as non-communicating sectors. The number of organisations concerned with resource use and environmental change, worldwide, must be very large and itself consume a great deal of energy and materials, especially paper, which we may hope to be recycled. What is extremely important, however, is the provision of widely available information. As ever, the possession of good information is a key to power and all the players in these manoeuvres know it. So there is still much secrecy in the whole field of resource use and environmental alteration, especially on the part of businesses and governments. If nothing else, it confirms the importance of the whole matter.

11

Strategies for global reform

Reform not revolution

Certain of the processes discussed so far in this book are identified as problems. If we were to construct a global agenda of resource-environment systems which require attention then it would almost certainly include:

- global warming, having ramifications into ozone depletion and carbon cycles, with the latter taking in deforestation as well as the combustion of fossil fuels.

- toxic materials, both those concentrated as industrial wastes and those dispersed as e.g. biocides.

- the delivery of nutrition and energy (especially fuelwood) in developing countries and the environmental degradation which accompanies shortfalls.

There are two basic kinds of reaction to this agenda (and the more complicated versions of it which can easily be compiled), which we can label (a) revolutionary, and (b) reforming. The former is one of the subjects of the next chapter and, as its name implies, thinks that only wholesale abandonment of present ways of doing things will produce any improvement. This applies especially to institutions such as governments. The latter argues that gradual reform is always more successful than revolution in the long run and that reform is more likely to carry people along than sudden change. Some forms of it as applied to institutions such as governments (as distinct from the freely-adopted behaviour of individuals) are discussed here.

Global-scale appraisals

The 1980s have seen a number of prestigious bodies enter the fray of providing world-level prescriptions for the future use of resources and for the scale and locations of consequent environmental impact. Some of these bodies saw it as an extension of their normal work (e.g. UNESCO, IUCN, WWF), and others (the Brandt and Brundtland Commissions) were specially convened groups.

Perhaps the highest profile of any of the reports that were published was obtained by the Brandt Commission (1980), whose chairman was Willy Brandt, Chancellor of the BRD 1969–74. Its main focus was development, as suggested by the title *North-South: a programme for survival.* The core issue was the alleviation of the material conditions of about 800 million people who comprise the poorest 40% of the South's population. Most of these are in countries where over two-thirds of the people are agriculturalists and where the economy is characterised by the export of large quantities of raw materials. Brandt also draws attention to the potentially precarious situation of some newly industrialising countries (NICs) which depend upon demand from the North, upon technology from the North and are in debt to the North. However, the basic needs of the poor in terms of health, housing and education (and especially of a supply of safe water, to which 4/5 of the South's population have no access) are at the core of the Brandt proposals. These centre round a programme of structural and ecological change (encompassing both emergency and long-term measures) in the major poverty belts of the world. The core programmes are mostly regional soil and water use projects, health care and the eradication of development diseases, afforestation, the use of solar energy, more exploration for minerals and oil, and improvements in infrastructural elements such as transport. These are, in effect, the delivery sites of global changes which depend upon the ability of the world-wide trade and financial organisations to bring about the large-scale transfer of money to LIEs. Interestingly, Brandt makes no mention of the possibility of negative results of any of these proposals, such as the global warming that might result from the use of more oil. In general, this report concentrates more on resources than on environment though it does state that ecological stability in LDCs is a necessary base for better conditions.

The report prepared under the direction of Gro Harlem Brundtland (1987), then Prime Minister of Norway, was commissioned by the UN General Assembly and the working group was explicitly labelled the 'World Commission on Environment and Development'. This starts from the proposition that 'the earth is one but the world is not' and hence adopts a perspective that flows from the nature of the global environment. To that extent it is environmentally deterministic, rather in the way that Brandt placed most emphasis on human institutions. Three groups of problems are identified by Brundtland: those of environment and resources, of sustainable development in LIEs, and of gaps in the institutional frameworks. To deal with these areas of difficulty, Brundtland erects a systematic series of analyses of population and human resources, food security, species and ecosytems, energy choices, the efficiency of industry, and the challenge of urbanism. The outcome of these examinations is a series of proposals for legal and institutional change at all levels from the UNO to national governments and also the role of the NGOs. Since it builds upon Brandt (and also upon Olaf Palme's analogous investigation of the prospects for international security), Brundtland is more thorough and wide-ranging, especially in its recognition of global environmental interdependence. However, critics have called attention to whether such features of its message as a 5–10 fold increase in world industrial output, and its ambivalence over biotechnology, and the status of Antarctica, are really a contribution to thinking about sustainability.

More limited than either of these, but more detailed in its particular sphere, was the joint IUCN, WWF and UNEP (1980) publication, the *World Conservation Strategy*, subtitled 'Living resource conservation for sustainable development'. This focuses initially upon ecosystems as life-support systems, the preservation of genetic diversity, and the sustainable utilisation of species and ecosystems. The World Conservation Strategy proposes national programmes for environmental protection in zones such as the TMFs, drylands, the oceans, the atmosphere and Antarctica, more North-South assistance for living resource conservation and stronger backing for international measures on conservation of species, and regional programmes for the protection of internationally shared seas and river basins. Each nation was supposed to prepare a response and a second version to appear which took into account these views. Explicitly, the World Conservation Strategy does not interface with the economics and ecology of industrialism, for example, but is often very much more detailed and realistic on the likely limits to ecological change in various regions.

In 1982, UNESCO brought together some of the findings of its programmes such as the IGCP, IHD, IOC and, especially, MAB (Man and the Biosphere). The integrated programme covered almost all aspects of human-planet interaction, from geological resources through natural hazards to environmental education and information. Much of it seems rather repetitive of the other documents and its chief strengths are in measures like World Heritage designations which confer a high profile on both natural and cultural monuments. However, the potential input of a UN *Cultural* organisation which might focus on environmental values and attitudes, and on cultural constructions of environment such as economics, seems to have been lost.

If anything is to come of these high-profile (and doubtless high-cost) evaluations then much of their impact must be at the level of national governments; their role in the present context must now be discussed.

National approaches

Nested within the global position, therefore, national governments have the choice of formulating environmental policies or failing to do so. If they have a national policy then its implementation may be partially pushed down to lower echelons of government; if there is no national policy then lower levels may in democratic societies try to work out one for themselves. Thus in the USA there may be no national land use strategy, but that does not prevent states and cities from adopting zoning procedures which control certain kinds of development. By contrast, in the UK the local authorities are responsible for enforcing national laws and regulations on development control, possibly irrespective of local conditions. In free-market economies there is always the presumption that any such bureaucratic control over environment should be kept to a minimum whereas in regulated, centrally-planned economies, the paramount role of the state is accepted.

Both the World Conservation Strategy and Brundtland asked for national responses, and these were prepared by both LIE and HIE countries. In the UK, for example, the response to the World Conservation Strategy appeared in 1983 and set out a generally predictable list of recommendations in terms of Great Britain's stance at home and overseas. Apart from the ritual call for better energy conservation, the major novelty was a suggestion for a Centre for a National Conservation Strategy, to which the response has been less than enthusiastic. The 1989 response to Brundtland follows much the same lines, but with the greater awareness since 1983, of global dimensions such as the models of climatic change. Once again, energy efficiency and conservation is seen as a vital ingredient. But neither response actually addresses itself to questions of sustainability. The definitions used ('Sustainable development means using man-made and natural resources in such a way that future generations are not left worse off' [1989]) are rather vaguely formulated, for example, and the focus on the conservation of species and ecosystems is restricted to seal pups and elephants.

More common as the basis for action has been the whole gamut of regional development plans which many countries, rich and poor, have adopted since such ideas became popular in the post-1945 period (Plate 3.5). A map of France in the 1970s, for example (Fig 3.3), shows the large number of areas subject to spatial management of one type or another.

Plate 3.5 New urban-industrial development under closely controlled conditions. These aim at producing a pleasant environment (a kind of parkland in this case) and of reducing the transference of land from other, less intensive, uses. Milton Keynes, England.

To the extent that these policies, complemented by control policies for certain types of protected areas, change land use then *de facto* they become the vectors of environmental management and change. If they protect (as in National Parks and areas of high landscape value, for instance), then they freeze an existing resource system in place; if they allow change then they provide the conditions for new systems which can be either creative or destructive in environmental terms. There is, however, no simple way of defining 'destructive' in this context: very pure environmental activists probably would want urbanisation thus classified but such a view is simplistic. At the very least, a compact city probably leaves more near-natural ecosystems intact outside its limits than an uncontrolled and piecemeal development of the same amount of housing and industry. At best it may house people whose thinking is the seed-bed of a better future, for cities have always been foci of innovative thinking and behaviour.

What future for planning?

The basic question in the issue of reform centres on the balance of legally-based regulation and economics-based *laissez-faire*. To achieve economies based on more sustainable resource and environmental foundations, seems very much like a job for central control: mutual coercion mutually agreed upon, might be one way of putting it. Yet in many nations, central planning is in retreat: witness the examples in the 1980s of the reaction to

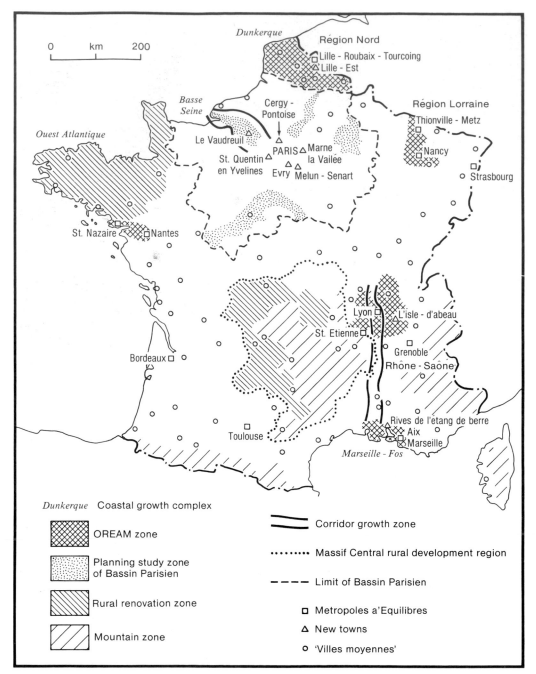

Fig. 3.3 Spatial management policies in France in the 1970s. A remarkable diversity of plans and policies is revealed. Source: J. V. Punter, 'France' in *Planning control in Europe*. London: DoE/HMSO, 1989.

Plate 3.6 Collective influence upon world events and processes: the UN Security Council voting during 1990. These votes led to the Gulf War, with its immense resource and environmental consequences.

it in Eastern Europe as well as the flight from mild forms of socialism in some liberal democracies. In the LIEs, the degree of control and infrastructure necessary to implement any central planning is often absent and indeed there is a school of thought which locates 'underdevelopment' precisely in ineffective government.

Where there is an ideological preference for market-led arrangements, the role of government is still important since no market will voluntarily accept the responsibility of dealing with external costs. These most often deal with wastes: the producer of soot from a chimney does not want to internalise the costs of the extra washing of clothing needed by those who live downwind from his factory, or the costs of medical treatment for those who contract respiratory diseases. Government regulations are still needed to enforce the recognition of the need to internalise costs and also to find suitable prices. If, for example, the costs of building a new motorway are to include the value of landscapes which are lost, who else is to put a price on a grove of ash trees or a group of old buildings? Asking the road-builders to do so is rather like allowing lions to specify the maximum speeds at which antelopes may be allowed to run.

However, most governments will acknowledge that they have a responsibility for 'the environment', if not for resources which in general they are willing to leave to the play of market forces as modified by some regulation. Governments cannot however avoid being involved at the international level and so the 1980s in particular have seen a growing number of international agreements which deal with environmental concerns (Plate 3.6).

These agreements are lodged in a particular place or places (or with the UN) for signature

Table 3.1   Global conventions on environmental protection, 1988: place, date and signatures in 1987

Conservation			
Wetlands		1971	49
World Heritage	Paris	1972	92
Endangered Species (CITES)	Washington	1975	96
Migratory Species	Iran	1979	28 + EC
Anti-Pollution			
Ocean Dumping	Mexico London Washington	1972	63
Pollution from ships (MARPOL)	London	1978	53
Biological Toxins	London Washington Moscow	1972	104
The ozone layers	Vienna	1985	35 + EC
Early notification of nuclear accident	Vienna	1986	31 + WHO
Ozone-depleting substances	Montreal	1987	23 + EC
Resource Use			
Antarctic marine resources	Canberra	1980	20 + EC
Law of the Sea	Montego Bay	1982	43 + EC
Tropical Timber Agreement	Geneva	1983	43 + EC
Antarctica minerals	Wellington	1988	6

Compiled from various sources

and the number of accessing states is a rough measure of global applicability. Since the first, in 1921, there have been about 140 of these but most are bi- or multi-lateral. The main globally important agreements since 1970 are given in Table 3.1. They deal with three main areas:

- *The conservation of species and ecosystems*   Here the importance for example of wetlands to migratory waterfowl is recognised in the Ramsar Convention of 1971, and in the convention on international trade in endangered species (CITES), lodged in Washington in 1973.

- *Anti-pollution measures*   A great impetus to this area was given in 1963 by the Treaty which banned the atmospheric testing of nuclear weapons, and several measures have failed which address both regional and global problems. The latest have been aimed at the risk of global warming and other results of the depletion of the ozone layer of the atmosphere and more effort in this field is likely, especially over the question of carbon dioxide emissions.

- *Resource use*   The clearing of the TMFs has occasioned a mild approach to the problem via the International Tropical Timber Agreement, and the Antarctic resource use conventions have tried to formulate policies for the exploitation of the region without damaging the environmental systems. The Law of the Sea convention, deposited in Montego Bay (Jamaica) in 1982, but not yet in force, must rank as one of the most powerful and far-reaching of international agreements on resource use.

Several of these agreements have spawned special organisations, like the International Tropical Timber Organisation (ITTO), established at Yokohama (Japan) in 1987. Most, however, rely on national governments to enact domestic legislation to ensure that they conform to the treaties and conventions. The 1987 Montreal Protocol on ozone-depleting substances is thought to be a pointer to the future since not only did it require rapid action in the shape of a 50% reduction of CFC production by 1999 but recognised the position of the developing countries by allowing them increases in CFC use so as not to retard the use of refrigeration with its beneficial effects on health.

12

Revolution better than evolution?

Alternative futures?

One scenario for the future can be summed up as 'business as usual'. This is basically the idea that present ways of managing our resources and environment will continue. The conservatism of individuals and institutions will make this true, though some reform (along the lines of Ch 11) will take place more likely than not, in order to accommodate perceived new realities about the nature of the environment.

Nevertheless given the restless and explorative nature of *Homo sapiens*, it seems unrealistic to think that there will not be more technology and more complex mechanisms, so long as there is the energy to make and run them. This might lead us towards a *technocentric* future. The replacement of all the support and feedback mechanisms of the natural flows of the planet by technological devices is rather unlikely, but it might make sense to try and decouple the human economy from that of nature. There are some limits of this in plants to build self-contained cities on offshore islands, or to put power stations (usually unclear) similarly offshore or deep underground. Renewed interest in relatively 'tight' cycles of non-renewable minerals points in roughly the same direction. At its extreme, such a scenario would presumably encase the 'econosphere' under plexiglass domes so that all materials, including gases, were kept away from the natural biogeo-chemical cycles. If the human population could be fed then the lands and seas (and atmosphere) outside could be allowed to revert to a condition of minimum impact. It sounds like SciFi (and could end up more like the city of the movie *Blade Runner* than Utopia) but in 1800 who would have thought of Disneyworld?

But it is also our task here, at the end of the book, to explore in a preliminary way whether there are other paths available to the human species. To ask, for example, whether revolutionary behaviour is likely in our environmental relations, as distinct from piecemeal shifts towards reform or indeed as different from a *laissez-faire* fatalism. And also to ask whether we have anything to learn from the history and progress of pre-human nature, which was so successful at filling the ecosphere with so many forms of complex life.

Many writers in the field of resources and environment disavow the likelihood of revolution, i.e. of sudden changes in behaviour which spring from the people themselves and not from their governing institutions. They point to the events in Eastern Europe in 1989, for instance, and note that while environmental concerns were painted on some of the banners of the revolutionaries, they were subordinate to the major political aim of being told the truth. Nevertheless, while environmental concerns are being taken so seriously at present (and more so than in the last 'environmental crisis' of 1965–72), it is appropriate to look, though in no great detail, at some of the radically different ways of constructing the humanity-nature relationship.

Holistic geophysiology

This is a complex term for a relatively simple idea, namely that the planet's parts and linkages are indeed one whole. The basic concept has been given impetus by the Gaia hypothesis (named after the Greek goddess of the earth) of J.R. Lovelock, in which he suggests that much of the physics and chemistry of Earth works so as to maximise the conditions for life, noting for example that the salinity of the oceans and the gaseous composition of the atmosphere would be different had life not existed (Plate 3.7). So, to some extent these inorganic parts of the globe are extensions of its living organisms and systems. A cat's fur and a wasp's nest are necessary for the survival of the cat and the wasp but are not themselves living. The hypothesis predicts that life should, for example, affect climate. The discovery that oceanic phytoplankton emit dimethyl sulphide (DMS) which as aerosol droplets form the condensation nuclei for rainfall seems to confirm this. So the photosynthetic rate of phytoplankton affects the rainfall over the continents.

The consequences for humans are profound. We are dethroned in the sense that a greater entity is maximising the conditions for life but not necessarily our life; presumably if we transgress the Gaian conditions then our tenure will be short. Hence we have to know which of our activities are most likely to lead to the type of system perturbations which we might not survive, and rapid and fluctuating climatic change seems to be the best candidate so far. In this, the role of living organisms in fixing carbon seems very

Plate 3.7 Individual influence: J.R. Lovelock FRS, whose formulation of the 'Gaia hypothesis' has had a considerable effect upon scientific and humanistic thinking about the planet Earth and the human position within her processes.

important, so that if we fell large areas of rapidly-growing forest and poison the tropical seas with the fastest productivities of phytoplankton, then we will be hastening our own demise.

In an even wider and longer perspective, Gaia is of course dependent upon energy from the sun. When that has gone (which will take rather longer than the next set of examinations) then life will disappear. Life itself cannot escape this bounding condition but can as it were intervene temporarily in the flow of energy from sun to Earth and back to space in order to build up complex structures like plants, animals, fungi and bacteria. These complex structures take in energy and then dissipate it as heat but do something remarkable in the process, like making possible the evolution of tropical orchids and the whale, for example. This knowledge, too, gives us a perspective upon our own behaviour. Since the supply of energy from the sun which is able to facilitate these processes is limited (whether as fossil fuel stores, fixed as chemical energy by plants, or dispersed as 'unused' sunlight) then we should act so as to keep to a minimum the rate of dissipation of this highly potent energy into heat which can accomplish nothing.

The economist Georgescu-Roegen has started from this evaluation of energy flows and formulated a programme of economic response. In this, the first requirement is to abolish the manufacture of the instruments of war and then to outlaw war itself (we are talking radical here, let us remind ourselves). Then, all economic processes should be audited for thermodynamic thrift: they should consume the minimum amounts of energy. This leads to such matters as increased rates of re-use, repair and recycling of materials, allied to goods that lasted much longer. Shifts in demand would be needed, including the sovereignty in the West of changes in fashion every few months for cars or clothing, for example. Thus the Volvo car and the calico shirt are the models for the future. Extension of the ideas brings in the precept that the global population should eventually be of a size that can be supported solely upon organic farming.

This 'bioeconomic' programme conflicts little if at all with Gaian ideas. They both have in common, we ought to note, a sense of the precariousness of the position of the human species. Georgescu-Roegen even suggests that our time on earth might be quite a limited one and that we are destined for (in evolutionary terms) a short life but a merry one and have no justification for assuming that *Homo* is here to stay.

No environmental problems

One lesson from these statements is that there is nothing wrong with the planet as such: we are the problem. There is no such thing as an environment problem, that is to say, only human problems with an environmental delivery route. We must look to our social and economic practices first. The main targets for close scrutiny seem to be processes which poison ecological systems (both natural and human-directed), which also introduce instability into such systems and hence unmanageable rates of change. There are two obvious groups who engage in these resilience-reducing transformations:

- The high-consumption (i.e. of materials and energy) societies of the North, who cause the manipulation of many systems on a world-wide basis by creating a very high demand for resources and then disposing of high quantities of wastes, some of which are substances unknown in nature. Security of supply is important and merits large-scale investment against the future.

- The poorest groups in the LIEs, who have no choice but to deplete the resources of their immediate environment. They must clear a forest or a slope to grow crops, fell trees to get firewood or graze large herds against the likelihood of losses from drought in order to survive. Children are the only worthwhile investment in these insecure conditions.

Two aspects of these categories need more examination: (a) are these conditions linked in any way, and (b) what drives the first group, given that the second is forced by simple necessity? Proposition (a) will be discussed on p 232–4; the second proposition gets some introspection next.

Deep ecology

This label is given to a movement which is largely inspired by the writings of a Norwegian philosopher, Arne Naess, born in 1912. The movement, which is not organised in any way, rejects reform as being an insufficient response to the current environmental situation. The fight against pollution and resource depletion, for example, is seen as 'shallow ecology' aimed largely at perpetuating the affluence of people in the HIEs. By contrast, Deep Ecology rejects the dualism of 'man-in-environment' in favour of a total egalitarianism of the biosphere, with all the changes in perception and relationships thus implied (Plate 3.8). An equal right to live and bloom is accorded to all life, along with the knowledge that killing, for example, is part of any normal ecological pattern. This contrasts with the usual

Plate 3.8   Direct intervention to protect the natural world. Members of the Chipko movement (Karnatka, India, 1989) embrace trees to try to prevent them from being felled. Note that this woodland has clearly been managed as a renewable resource.

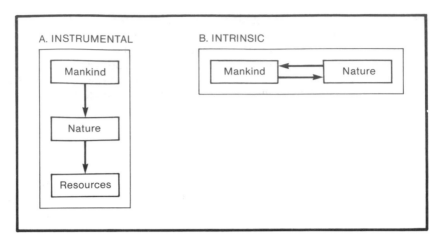

Fig. 3.4 Valuations of nature in two worlds: (A) with a hierarchy of dominance, and (B) with equal value accorded both parts within the whole.

instrumental Western worldview and is centred upon the idea of the intrinsic worth of the non-human (Fig 3.4).

These basic ideas have been formulated into a 'platform', which contains the following elaborations:

- The intrinsic value of both human and non-human life, with the latter possessing value independent of their usefulness for human purposes.

- Diversity of life forms is valuable in itself and humans have no right to reduce this except to satisfy basic and vital needs.

- To reduce the impact of humans upon the rest of the world and to ensure the flourishing of human lives and cultures, a decrease in the number of humans is needed.

- Significant changes in policies of all kinds are needed in the direction of appreciating life quality. This latter is more important than high material standards and requires pleasure and satisfaction to be gained from a close partnership with other forms of life.

- An obligation to participate in one way or another in an attempt to implement the necessary changes.

The difficult part, conceptually, in Deep Ecology is the notion of the *realisation of the self*, which is concerned with the conditions under which each one of us flourishes ('blossoms' in Naess's words) to the maximum extend as an individual. Proponents of Deep Ecology argue that in the West, our flowering has become defined in terms of material possessions and throughput, just as we have used data for energy consumption in this book. They suggest, as did Spinoza (1632–77), that the narrow ego of a small child needs to be developed into a comprehensive structure that comprises all human beings. For the Deep Ecology movement, this is insufficient: the comprehensiveness must extent to an identification with all life-forms. 'Environment' (which disappears as such) is not something 'outside' which has to be adapted to, but something valuable which we want to treat with respect and whose richness we may use to satisfy our vital needs. Such a maturity guarantees action which may be called beautiful since it results from a wholeness of personality and of nature.

It will surprise nobody that Deep Ecology has its detractors. Most of these ignore it completely, as being irrelevant to the 'realities' of today, especially the poor in the LDCs; others argue against its model of the equality of all living things; yet others reject its notions of self. However, there is no doubt that it brings together concepts about what it means to be human and of the human place in the world in a way that is specifically addressed to today's perceived problems. Such concerns have also been examined in other systems of philosophical and religious thought.

Other approaches to the self

This discussion has started with Deep Ecology because it is completely opposite to the Western worldview which currently predominates and which has an *instrumental* attitude to nature, i.e. values everything in terms of its usefulness to human society. But there is a long tradition of seeking to define the self in terms other than those of material well-being and we should see if any of them are relevant to our evaluation of alternatives.

Western religions

In the sense that Judaism is at the foundation of the world-view of only one nation whereas many espouse (not always explicitly) a Christian outlook, it is the latter religion which ought to have a developed stance on the relation of mankind and nature. In the last 20 years, there has been quite a vigorous debate about the stance of Christianity, ever since the historian Lynn White developed his thesis that in the early words of *Genesis* (I: 26–29) there lay the foundations of an exploitive attitude to the natural world. Being fruitful, multiplying and having dominion over the earth was seen as an inevitable consequence of Man being the only creature made explicitly in God's image. The ripostes to this somewhat vigorous reinterpretation of the Scriptures were various. One class centred around the notion of mankind as Stewards of an estate, which they have a duty to hand back to the Owner in as good a condition as when they took it over. Others have stressed an incarnational approach in which God has been, is and will be in all things: a condition known as *pan-en-theism*. Most recently, theologians have taken more of a turn towards the celebratory, invoking wonder as a basis for a co-evolutionary relationship between humans and everything else as co-creators of a *cosmos*, when that word is taken to mean a world with order, i.e. not a chaos.

Christianity has had interpreters who could fit comfortably alongside an instrumental world-view: a recent example is the Encyclical *Laborem Exercens* of Pope John Paul II, where the forcing of nature into productivity for human ends is seen as a kind of quantitative measure of the grandeur of mankind. But by contrast there are such historic figures as Hildegard of Bingen (1098–1179) and Francis of Assisi (1181–1226) for whom a basic spiritual equality of all things is a starting-point and who thus appeal to those of a deeper Green outlook: indeed, Hildegard's coining of '*viriditas*' ('Green Truth') would have earned her convent many a Deutschmark had the Berne Convention on copyright been in existence in her time (Plate 3.9).

Other religions

Since the 18th century, Christianity has been more important than any other religion because it has been so tied in with a world-view which has dominated the world via economic and political hegemony. Other religions are of course extant, but seem to exert little influence on resource-environment systems except at local levels. Nevertheless, they

Plate 3.9 A reclaimable resource? The medieval nun Hildegard of Bingen, author of a great deal of 'Green' poetry which appears to be prescient in its view of the degradation of nature. Such writing reminds us of the heritage of thought about, and celebration of, nature which may form a resource for the present.

are sometimes seen as a seed-bed of different attitudes which could be added to the Western worldview to its great advantage.

The North American aboriginal populations, for example, nearly all held that the people occupied a sacred space. All their actions, including those of subsistence, needed sanction from a god or gods. Thus to kill to eat was permissible but was part of a whole cosmos. These religions have been submerged under the weight of European colonisation but are sometimes held up as examples of a more tender approach to the land than is customary in the USA, especially under Republican administrations. North Americans, too, were sometimes influenced in the 1960s and 1970s by versions of Buddhism, notably the Japanese version much influenced by Chinese Taoism known as Zen. The Tao maintained a quietistic attitude to life: harmony with the cosmos was to be sought by finding its ways and rhythms and adapting to them rather than trying to alter things and other people. Zen's main contribution has been in stressing the unity of all things and the primacy of experiential knowledge over objective rationality. Enlightenment comes suddenly and inconsequentially but an ascetic regime is often helpful. Thus Zen leads to a low-impact life-style. In this it is typical of mainstream Buddhism, where the environment can be an object of attachment and therefore of suffering. Thus any devotion to worldly things that derive from the environment (e.g. material possessions) will lead to unhappiness and the soul will not escape from the continual process of re-birth. Detachment usually means as well a low-impact life-style which precludes the taking of life and leads to vegetarianism.

In Hindu cultures, there is a long tradition of environmental protection known as non-injury or *ahimsa*. The life-style advocated by the Mahatma Ghandi constitutes a predisposition to a low environmental impact, though in areas of marginal rainfall, Hindus have scarcely been exempt from problems of land and water use. However, Hinduism does not seem very capable of export to the West, compared with Buddhism for example.

Islam is of course a proselytising religion, although few of its more permanent conquests have been west of its heartland. The Holy Quoran is explicit that the earth is only a temporary home, even though man is a superior being. Humans are stewards of the gifts of Allah, and must administer them with justice, piety and a due knowledge of scientific and environmental matters. The discovery of oil and more liberal interpretations of Islam in some Middle Eastern countries have allowed development of a basically Western character. It remains to be seen whether Islamic fundamentalism of the Iranian pattern will produce markedly different patterns of resource and environmental usage.

In both East and West, most developed religions (as distinct from folk beliefs) have collaborated in human behaviour that is destructive of species, habitat and system, and with non-sustainable development. Some kind of syncretic faith with elements of all the major religions might help with the development of a world-view which is more tender to nature. Yet religions, like languages, seem resistant to recombination into structured forms: think of the example of Esperanto. If they are to play a part in any new relationships between humans and the natural world, then it is more likely to come through reconstructions of the historic faiths.

Western meta-narratives

Some apparently recherché stretches of the Western philosophical tradition may seem an inappropriate inclusion in this book. But it is clear that behind all the patterns of resource use and environmental impact there are systems of thought which sanction all the various actions which societies and individuals take. It is not surprising therefore that scholars should try to develop theories: systems of explanation which apply in an over-riding fashion to all cases through space and time. Theories of the relations between humans, technology and the environment are not in short supply but most of them have been transferred wholesale from either the natural or the social sciences. Robert Kates has set out examples of three sorts of theory which deal with this area of human concern:

- *One-dimensional causality* Here there is a simple A causes B approach. Something is the overall explanation of everything: God; the natural environment as a determinant of human affairs; and technology once let loose like a genie from a bottle, are examples.

- *Partial theories* These are dominated by the Malthusian idea that it is the equation of population and resources which drives all the interactions which have been our subject matter. The idea of the global commons in which every individual appropriation of space or resource diminishes the patrimony of the rest is strong here; the work of Karl Marx in linking productive forces to social relationships is also central.

- *Interactive theories* Good examples of these are accounting systems built around, for example, energy: in the way this has been used here to link resource processes and environment or elsewhere to periodise history. Attempts to put a price on everything in environmental decision-making so as to reflect true costs of development projects are analogous. Systems models which use computers to model everything that can be measured are also in this category.

Theories can be used in a purely descriptive sense, i.e. to bring order and explanation into observed patterns. The poorest nations in the world (Fig 3.5) do not appear to be randomly distributed and Kates tried to use theories to describe and account for the location of these 'least-developed' nations. Thus we could start with the characteristics of the nations themselves (newly independent in many cases, poorly endowed with most kinds of resources); or, theories of the process of underdevelopment in which the North in various ways exploits the South and keeps it poor; or, theories in which the very concept of development is seen as an inappropriate scale upon which to measure the cultural, economic and environmental condition of a nation. In the end, a combination seemed the most satisfactory: these places were marginal environmentally by virtue of isolation, relief or climate and marginal economically to the world's main flows of economic activity. Environment, it seems, does not create poverty but can exacerbate the position of the very poor.

An even more common use of theory is in a predictive and prescriptive sense, to provide an idea of what the future holds and if necessary redirect processes towards a more desirable future. Given the complexities with which we deal here, this seems possible only on the broadest of scales. For example, in looking towards a theory-guided perspective on the year 2048, Burton and Kates took as central that there would be 'a new relationship with the environment. Decisions would be guided by a concern for the long-term viability of the biosphere'.

It is thus interesting that in this field of study as in most others, the tendency of humans is to seek general explanations. Partly this seems to be as an intellectual satisfaction but more important, surely, is the urge towards control. If we can find a general law which controls all processes of a given type then we can construct our own future as we want it, providing we are the ones with the information and the means. Yet possession of means does not necessarily imply that the right ends will be sought: such is the task of justice, which must be briefly considered.

Justice and the environment

Justice and equity are concerned with fairness: they are essentially moral concepts which deal with the distribution of shares of anything. The principles upon which they are based are not, though, always the same. The conservative will say that the greatest shares should go to those who risk most, work hardest and most entrepreneurially and respect tradition. There will always be somebody at the bottom of the pile, however: that is how things are. The socialist will work for an equal share for everybody in need, enforced not by a market mechanism but by laws. Common to both, however, even in the most conservative of societies is the duty of humans to others (and perhaps to their environment) to see that basic necessities are supplied to all, whether by regulation or by charity.

It is clear that there is some inequity in the way the world uses its resources and manipulates its environment, and we have wondered if there is any relation between the rich and their choice-led metamorphoses of environment, and the poor with little if any

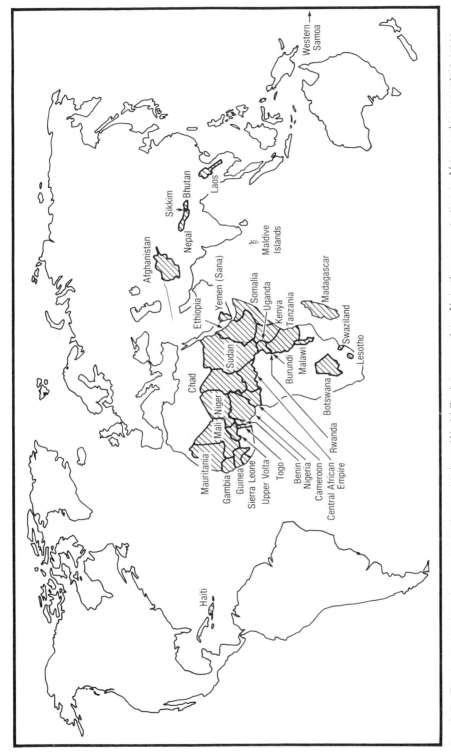

Fig. 3.5 The least developed countries in the world, from World Bank income data. Note the concentration in Africa and in mountain areas. Source: R. W. Kates, 'Theories of nature, society and technology', in E. Baark and U. Svedin (eds) *Man, nature and technology*. London: Macmillan, 1988.

choice. While a small proportion of the population uses the majority of the resources, it is always arguable that the minority group is keeping its privileged position by making sure that the LDCs are restricted in their access to technology, capital and knowledge. Only debt seems to be in unlimited supply. The opposite view is that it is only the demand exerted by industrialised nations which provides a basis for many LIE economies and so any lower levels of industrialisation in the Western nations will result in even lower material standards in the South.

The balance of the argument seems to rest with the views of the Brandt and Brundtland reports (see p 217): export-led economies seem especially prone to distortions of equity whereas self-sufficiency seems to provide better conditions for a wider group of people, as well as necessitating an approach to the sustained use of renewable resources which is compatible with a lower environmental impact. So more self-sufficiency and lower demand from DCs seem fundamental; lest this is thought too radical, most of the industrial group could very likely support themselves on recycling and re-use together with higher levels of energy efficiency. Ways already mentioned of trading debts for better policies leading towards ecological conservation and sustainable development seem to be a hopeful part of such a package.

In most civilised places, poverty is seen as an affront to justice. And it appears that 25% (675 million) of the people of Asia, 62% (325 million) in sub-Saharan Africa, 35% (150 million) in Latin America, and 28% (75 million) in North Africa and the Middle East, are living in absolute poverty, and the rate of recruitment to that group is rising, mostly in the rural areas where 4/5 of the really poor live. Life is differentially harsh, too, for women and for children. The 'poverty trap' in which these and countless millions slightly better off than the absolute poor has three major teeth. The first is local, with poor access to land and food, and powerlessness against corruption. The second is national, where many policies on taxes and investment either ignore or militate against the poor. The third is global, where debt and falling export prices help nobody and least of all those who are most disadvantaged. Not only are poor people resource-starved but their condition forces them into practices which cause environmental degradation. In a complex set of interactions, landlessness is probably the worst problem, with the poor increasingly forced into marginal lands unable to support them. One calculation suggests that 47% of the world's absolute poor live on marginal or fragile lands. Getting out of the poverty trap is the subject of many other books but the linkage with the present theme undoubtedly seems to be through restoring access and control of resources to the very poor, using Gandhi's touchstone of the face of the poorest and weakest man. Will the change you contemplate, 'restore him to a control over his life and destiny?'

Freedom from want can best be pursued in an atmosphere of freedom from fear. The latter can come in packages of all sizes, from a massive thermonuclear war to another devastating circular from the Universities' Funding Council. But war is the most massive distortion of justice: not only do the wrong people always get killed but the activity uses up huge quantities of resources (even in our present relatively peaceful times, about 6% of oil consumption worldwide is directly used for military purposes, and the world average of share of GNP that goes to armed forces is 6%, i.e. about $200/cap/yr) and consumes large fractions of the budgets of the LDCs. For example, 21 days' worth of the world's armaments expenditure would pay for the whole UN programme for clean water supply to developing countries. In 1985, Ethiopia spent 9% of its GNP on its military; Mozambique 7.4%; Syria 22.8%, Nicaragua 16.8%. And we are all at risk from one or another of the effects of even a medium-scale nuclear war, with its threats of the 'nuclear winter', enhanced UV-B levels, and radioactive fallout together with total social and economic dislocation. It is worth remembering that both resource supply and environ-mental integrity are beneficiaries of non-violent conflict resolution.

Contexts for complexity

We have to recognise that the nature of the relationships in which humans find themselves is complex. There is no simple solution to the problems we perceive, no single road to be taken. Our cognitions must extend, for instance, to the realisation of a number of dichotomies: we are both material and cultural, being products of billions of years of cosmic evolution but yet having a neocortex that enables us to gain control over material things. Again, we are both emotional and rational, and can easily emphasise in our behaviour the one rather than the other. Yet further, we are ever-changing, as are natural ecosystems, but have powerful constraints built into many of our value systems.

Each of these dichotomies can lead us into attitudes which are inimical to fruitful and creative coexistence with the non-human constituents of our planet and the rest of the material cosmos. Our role as cultural creators, for example, can lead us into forgetting that an evolving nature lays down conditions for all its living systems. Or it can lead us to an apathy in which we acknowledge our guilt at having 'screwed the place up' but fail the test of doing anything better. We can indeed 'think globally' but are paralysed from 'acting locally'.

What seems to be needed is affirmation of the acceptability of all these traits, plus the effort to bring them into a harmonious balance. The history of the West (and hence now of an almost universal worldview) has been to put these pairs of dichotomies into conflict so that conquest was always necessary: of ourselves, of other people with the 'wrong' outlook, and of nature. We need new worldviews, new metaphors (the views of Earth from space may be part of such programmes) and new ways of motivating us emotionally to do what we rationally acknowledge to be correct. We are aiming to replace the climbers' 'we have conquered Mt Everest' with 'we have made friends with Mt Everest'.

A piece of the main

For the more immediate future, the history of the last 10,000 years (and in reality probably that of the last 0.5 million years, ever since the genus *Homo* learned to control fire) leads us to suspect that models and theories about mankind-environment relationships which are based on the existence of an equilibrium state are unlikely to be either true or helpful in a practical sense. The reality is that humans have always upset equilibria, producing times of rapid change followed by new resting phases (which may look like equilibria in a short time-prespective) which are then followed by more change.

This is not totally unlike some relatively recent models of the progress of organic evolution of the type first elaborated by Charles Darwin and it makes us wonder whether some model of co-evolution of humanity and the rest of the planet is not a viable role-model for the future. If evolution consists in part of adaptation to the forces of a sur-rounding environment, then there is a stress cascade in which everything is subject to the biophysical systems of the planet even though living organisms (including our own species) try to adapt the physics and chemistry (especially if we believe the Gaia hypothesis) to their own continuation. Hence, humans become part of the evolutionary pressure upon the other inhabitants of the planet.

The signs are, though, that pressures generated by human societies are too intense when added to those of a short-lived interglacial period and that the rates of change produced in the 19th and 20th centuries are about to result in periods of change at rates which may well prove to be inimical not only to ordered human societies ('future shock') but to many other species and systems as well. After all, these are inseparable, since the ecosphere and the econosphere are at present coupled together, interchanging energy and materials.

The above suggests that it would be a prudent move to lessen the human-generated pressures in the global system. This would be for the good of the planet as a set of continuing resources for humanity if we take an instrumental view of it. If we allow some intrinsic value to the Earth, then the worth of this change of direction seems even greater. More controversial but needing very close study is the idea that lower pressures on resources and environment in the industrialised world would enable the LDC's to be more self-sufficient and also to reduce destabilising pressures on soils, water, living organisms and, critically the atmosphere. And, critically, we all need to see resources and environment as whole processes. There can be no discussion of supply without that of creation of wastes and no study of waste treatment without consideration of resource use and allocation.

The actions which have to be taken to re-orient us towards a more co-habitatory and eventually co-evolutionary stance must in part stem from the governments of nation states. To the extent that some of them are now developing more 'environmentally friendly' legislation (often under pressure from Green Party politics), they deserve support and encouragement and the recognition that they have to proceed slowly so as not to alienate their electorates. They must, above all, be encouraged to develop demand structures which encourage low-impact resource use, and even the most avid free-market enthusiasts can in their more sober moments see the need for that.

Beyond governments, however, there stand the people. In any system of government, even the most centralised, they can exert control (as was shown vividly in the events of 1989 in Eastern Europe) provided they have the information of the kind often provided nowadays by television about other events and processes. So the responsibility of the individual cannot be side-stepped and we have to undertake the painful process of deciding whether we wish to define ourselves as consumers of the earth or carers of it. Which, neatly, points to our role as 'consumers'. In the North we function thus unless we are members of an eco-commune or some other self-denying organisation. To become consumers is the aim of the people of many politically repressed nations of the world and of LIEs, not least because advertising, films and TV speak of the manifold benefits of being a consumer.

So it is no use telling anybody outside the rich nations that their ambitions are probably not capable of being fulfilled because the systems of the earth cannot stand the additional pressures thus generated. Only if the rich reduce their consumption is there likely to be progress; and only if individuals act will there be change since any elected government that actually promised a permanently lower material standard of living would not last the week. But individuals and families, and small communities can start: by refusing to consume at the rates which advertising tells them they should, for example, or refusing to buy goods which are packaged to a totally unnecessary level. We can take part in recycling schemes for all kinds of materials and refuse to accept that they cannot run because there are then market gluts of the used materials, such as paper, since places where governments demand a recycled content to all the paper they buy do not have such difficulties. Equally, we can buy goods that are repairable or re-usable for another purpose, following the example of many LDCs. Following the general theme of this book we need to move to be lower consumers of energy, either in its direct or embedded forms.

In terms of the light streaks with which we began the book, we need to think of the richer people having the brightness of their connectivities diminished and the length of the streaks shortened. More colours are discovered (for lower quantities) and have shorter and more circular trajectories (for re-use and recycling). The poorer people, by contrast, need to have brighter lights (literally and metaphorically) but not to be so dependent on the streaks coming a long distance.

Because the rich nations draw in their resources from so many places, the act of using less and creating fewer wastes links us all in the most obvious way to the rest of the human-directed and natural systems of the entire planet: by acting locally, we think globally. In my opinion, one of the extremely important developing strands in this tying together (with a potential for producing change of many kinds in any direction) is that of satellite TV, which will produce the 'global village' in a way never before dreamt of (Plate 3.10). (I stress

Plate 3.10 The global village. The Mursi in Ethiopia watching TV, the pictures of which could, by satellite, be an instantaneous relay of events anywhere in the world.

TV because it seems to be the validating medium for many cultures in the way that newspapers were in the West of the 1920–50 period and the Christian Church was in mediaeval Europe.) It can disseminate global consumerism of the Western type. It could also disseminate the need for global ecological knowledge, a respect for natural and cultural diversity, a sense of global responsibility and caring. In social terms, it could foster life-styles based on non-violence and post-patriarchal values and revive the value of the small unit at work, at play and for living among. All these aims which focus eventually upon life being a matter of quality rather than quantity could be stressed: it remains to be seen whether this is so or whether they are suppressed in favour of global game shows and chats with those people who are famous only for being famous, for 15 minutes at least.

The heading of this last section is from John Donne's most often quoted piece (*Devotions*, XVII), which begins

> No man is an island, entire of
> itself; every man is a piece of
> the Continent, a part of the
> main.

which has become true, especially since the 19th century, in a way undreamt of between 1571 and 1631. As we know, Donne goes on to suggest that every person's death therefore involves everybody else, because they are involved in mankind. Famously, it ends

> And therefore never send to
> know for whom the bell tolls;
> it tolls for thee.

But we are now involved in much more than mankind: we are involved with the whole Earth and its denizens and materials. Given our knowledge and our technology, we might be able to convert the whole planet to a gigantic people-food-structures system which might be high on stability but distinctly low on variety; for many this would indeed be a death of a sort. When the potential for life in a world with biological, scenic and cultural diversity is also present, is this not the better choice? George Bernard Shaw said that we must get what we like otherwise we will have to like what we get: why not make the bells ring for something we like rather than toll for something we have merely got?

Further reading

Regionalisations

Biswas, A.K. (ed) 1984: *Climate and development*. Dublin: Tycooly Press Natural Resources and the Environment vol 13.

Brown, L.R. and Wolf, E.C. 1987: 'Charting a sustainable course', in L.R. Brown (ed) *State of the world 1987*. New York and London: W.W. Norton, 196–214.

Chandler, W.U. 1987: 'Designing sustainable economies', in L.R. Brown (ed) *State of the world 1987*. New York and London: W.W. Norton, 177–95.

Durning, A.B. 1989: 'Mobilising at the grass roots', in L.R. Brown (ed) *State of the world 1989*. New York and London: W.W. Norton, 154–73.

Eyre, S.R. 1979: *The real wealth of nations*. London: Edward Arnold.

Hansen, S. 1989: 'Debt for nature swaps – overview and discussion of key issues', *Ecological Economics* 1, 77–93.

Jacobson, J. 1988: 'Planning the global family', in L. R. Brown (ed) *State of the world 1988*. New York and London: W. W. Norton, 151–69.

Lapp, F. M. and Schurman, R. 1989: *Taking population seriously*. London: Earthscan Publications.

Pearce, D. 1988: 'The sustainable use of natural resources in developing countries', in R. K. Turner (ed) *Sustainable environmental management: principles and practice*. London: Bellhaven Press, 102–17.

Pearce, D. 1989: 'Sustainable futures: some economic issues', in D. Botkin *et al* (eds) *Changing the global environment. Perspectives on human involvement*. London etc: Academic Press, 311–23.

Redclift, M. 1984: *Development and the environmental crisis. Red or green alternatives?* London and New York: Methuen.

Redclift, M. 1987: *Sustainable development. Exploring the contradictions*. London and New York: Methuen.

Riddell, R. 1981: *Ecodevelopment*. Farnborough: Gower Publishing.

Thijs de la Court 1990: *Beyond Brundtland. Green development in the 1990's*. New York: New Horizons Press/London: Zed Books [first published in Dutch in 1988].

Frameworks

Berkes, F. (ed) 1989: *Common property resources: ecology and community-based sustainable development*. London: Bellhaven Press.

Berkes, F. Feeny, D. McCay B. J. and Acheson, J. M. 1989: 'The benefits of the commons', *Nature* 340, 91–92.

Hardin, G. 1968: 'The tragedy of the commons', *Science* 162, 1243–48.

Hardin, G. and Baden, J. (eds) 1977: *Managing the commons*. San Francisco: W. H. Freeman.

Kay, D. A. and Skolnikoff, E. B. 1972: *World eco-crisis. International groups in response*. Madison, Wis: Univ of Wisconsin Press, 1972.

Lowe, P. and Goyden, J. 1983: *Environmental groups in politics*. London: Allen and Unwin.

Shrybman, S. 1990: 'Free trade vs the environment: the implications of GATT', *The Ecologist* 20 (1) 30–34.

Strategies for reform

Brandt, W. (ed) 1980: *North-South: A programme for survival*. London and Sydney: Pan Books.

A. Cutrera, (ed) 1987: *European Environment Yearbook 1987*. London: DocTer Institute.

DoE [UK] 1989: *Planning control in western Europe*. London: HMSO.

DoE [UK] 1989: *Sustaining our common future*. London: DoE.

Gro Harlem Brundtland, (ed) 1987: *Our common future. World commission on environment and development*. Oxford and New York: OUP.

IUCN/UNEP/WWF 1980: *World conservation strategy. Living resource conservation for sustainable development*. Gland, Switzerland: IUCN.

Johnson, B. 1983: *The conservation and development programme for the UK. An overview. Britain*. London: Kogan Page.

Johnson, S. P. and Corcelle, E. 1989: *The Environmental policy of the European communities*. London: Graham and Trotman.

UNESCO 1982: 'The environment and resources: a six-year plan', *Nature and Resources* 18 (4) 18–24.

Revolution better than Reform?

Barnaby, F. 1991: 'The environmental impact of the Gulf war' *The Ecologist* 21, 166–72.

Black, J. 1970: *The dominion of man. The search for ecological responsibility.* Edinburgh University Press.

Brown, L. R. 1986: 'Redefining national security', in L. Starke (ed) *State of the world 1986.* New York and London: W. W. Norton, 195–211.

Burton, I. and R. W. Kates, 1986: 'The great climacteric 1798–2048', in R. W. Kates and I. Burton (eds) *Geography, resources and environment, vol II: Themes from the work of Gilbert F. White.* Chicago: The University Press, 339–60.

Cobb, J. 1983: *Ecology and religion: towards a new Christian theology of nature.* Ramsey, NJ: Paulist Press.

Dobson, A. 1990: *Green Political Thought.* London: Unwin Wyman.

Durning, A. B. 1989: Poverty and the Environment: Reversing the Downward Spiral. Washington DC: *Worldwatch Paper 92.*

Durning, A. B. 1990: Apartheid's Environmental Toll, Washington DC: *Worldwatch Paper* 95.

Engel, R. 1988: 'Ethics', in D. C. Pitt (ed) *The future of the environment. The social dimensions of conservation and ecological alternatives.* London and New York: Routledge, 23–45.

Kates, R. W. 1988: 'Theories of nature, science and technology', in E. Baark and U. Svedin (eds) *Man, nature and technology: essays on the role of ideological perceptions.* London: Macmillan, 7–36.

Kats, G. K. 1990: 'Slowing global warming and sustaining development'. The promise of energy efficiency, *Energy Policy*, **18**, 25–33.

Naess, A. (trans D. Rothenberg) 1989: *Ecology, Community and Lifestyle.* Cambridge University Press. (A translation of some parts of *Økologi, samfunn og livsstil.* Oslo: Universitetsforlaget, 1976).

Naess, A. 1988: 'Deep ecology and ultimate premises', *The Ecologist* 18, 128–31.

Peters, K. E. 1989: 'Humanity in nature: conserving yet creating', *Zygon* 24, 469–85.

Renner, M. 1989: National Security: The Economic and Environmental Dimensions. Washington DC: *Worldwatch Paper* 89.

Westing, A. H. 1986: *Global resources and environmental conflict: environmental factors in strategic policy and action.* OUP for SIPRI.

Westing, A. H. (ed) 1988: *Cultural Norms, war and the environment.* OUP for SIPRI/UNEP.

Whyte, L. 1976: 'The historical roots of our ecologic crisis', *Science* 155, 1203–7.

Index

Compiled by Jessica Rackham, B.A. and Oliver Rackham, B.A.

Note: Numbers in italics refer to captions.

Acknowledgements:

The Hutchison Library – plates 1.1, 1.2, 1.7, 2.7, 2.13, 2.15, 2.22, 3.10
Topham Picture Source – plates 2.23, 2.31
Popperfoto – plates 2.27, 2.30, 3.6
Lehtikuva Oy – plate 2.32
Panos Pictures – plates 1.5, 2.28, 3.1, 3.8
Impact Photos – plate 2.19 © Homer Sykes
The Creative Company – plate 3.5
Safeway Plc – plate 2.8
James Lovelock – plate 3.7
National Power – plate 2.2
ICI – plate 2.21
Otto Muller Verlag – plate 3.9
A Shell Photograph – plates 1.8, 2.6, 2.10
Genut – plate 2.14
RMC Group Plc – plate 1.3
Chorley and Handford – plate 3.2
U.S. Geological Survey – plate 2.1 (Photo by J. D. Griggs, U.S. Geological Survey)
Barnaby's Picture Library – plate 1.9
Komatsu – plate 2.24
USTTA – plate 2.26
Vector Technologies – plate 2.3
The J. Allan Cash Photolibrary – plate 2.18
Greenpeace Communications Ltd – plates 2.5, 2.12, 2.17, 2.29, 2.33, 3.4
The Environmental Picture Library – plates 1.4, 2.20, 2.25
Tadeu Caldas/Emerson College – plate 2.9
Forestry Commission – plate 2.11
Sea Fish Industry Authority – plate 2.16
Canadian High Commission – plate 1.6
G. Grebby – plate 3.3
Department of Energy, Washington – plate 2.4